APPLIED
COST ENGINEERING

COST ENGINEERING

A Series of Reference Books and Textbooks

Editors

FORREST D. CLARK
Project Controls Consulting, Inc.
Green Valley, Arizona

A. B. LORENZONI
Project Management Consultant
Applied Project Management Inc.
Charlottesville, Virginia

Additional Volumes in Preparation

APPLIED COST ENGINEERING

Second Edition, Revised and Expanded

FORREST D. CLARK
Project Controls Consulting, Inc.
Green Valley, Arizona

A.B. LORENZONI
Project Management Consultant
Applied Project Management Inc.
Charlottesville, Virginia

MARCEL DEKKER, INC.   New York and Basel

Library of Congress Cataloging in Publication Data

Clark, Forrest D., [date]
    Applied cost engineering. Second Edition, Revised
and Expanded
    (Cost engineering ; 8)
    Includes index.
    1. Costs, Industrial. 2. Cost control.
I. Lorenzoni, A. B., [date]. II. Title.
III. Series: Cost engineering (Marcel Dekker, Inc.) ;
no. 8.
TS167.C53   1985   658.1'552   84-23040
ISBN 0-8247-7264-4

TS
167
C53
1985

MARCEL DEKKER, INC.
270 Madison Avenue, New York, New York 10016

Current printing (last digit):
10 9 8 7 6 5 4 3 2 1

PRINTED IN THE UNITED STATES OF AMERICA

We dedicate this book to two long-time supporters of cost engineering who were always willing to type, proofread, and advise during the preparation of this book, and without whose encouragement we probably never would have completed this work—our wives, Catherine and Elaine.

PREFACE TO THE SECOND EDITION

Since the first edition of this book was published in 1978, the field of cost engineering has greatly expanded. The economic recession in many parts of the world, inflation, and high interest rates have tended to focus corporate management and government interest on the budgeting process, cost control, and the cost engineers. As a response to this management interest, new cost control techniques have been developed and old ones perfected. Therefore, the authors have concluded that it is time to update *Applied Cost Engineering* so that it continues to reflect the latest advancements in the field of cost engineering.

In addition to updating the original text, a complete new chapter on cost control of subcontracts has been added to address the needs of readers in countries where subcontracting is prevalent. Other new subjects include an analysis of overtime costs, what to do about rework, and how to handle back charges. Additional information has been provided on bulk material control, and on monitoring construction field labor overheads. The chapter on labor productivity and forecasting direct labor manhours has been expanded to reflect additional data and experience on recent projects. The chapter on contingency has been rewritten to further address this knotty estimating/cost control problem.

At the request of persons using *Applied Cost Engineering* as a text in teaching the subject of cost engineering, a number of case studies have been included. These case studies reflect real on-the-job situations that confront practicing cost engineers, and are designed to expand the students' understanding of the fundamental principles discussed in the text.

Forrest D. Clark
A. B. Lorenzoni

PREFACE TO THE FIRST EDITION

This book fulfills the need for an overview of two major areas of the relatively new and rapidly expanding field of cost engineering. The American Association of Cost Engineers (AACE) defines cost engineering as "that area of engineering practice where engineering judgement and experiences are utilized in the application of scientific principles and techniques to the problems of cost estimation, cost control, and profitability." Two of these three areas of cost engineering—cost estimating and cost control—will be the subject of this text. Profitability, or "engineering economics," is amply covered in other texts.

Cost engineering is in its infancy. As recently as two decades ago, the term "cost engineer" was not at all common and very few of those engaged in the field on a full-time basis had strong academic background or training. In fact, the AACE itself was only founded in 1956. However, the field of cost engineering has grown significantly in both numbers and stature in recent years. The AACE is some 5000 members strong and many other international professional cost organizations have since been founded. No doubt, the agonizing periods of spiraling costs over recent years have created a more cost-conscious management in industry and have added considerable impetus to the growth of cost engineering.

Both of the areas discussed in the text have a tremendous impact on world and national economics and, hence, on individuals, societies, and living standards. This is often not understood or tends to be underestimated. For example, virtually all "go or no-go" decisions on projects, whether they be $100,000 or $10 billion, are made on the basis of economics, which in turn relies on the availability and accuracy of a cost estimate. This applies to both private industry and governments. National budgets themselves are nothing more than a compilation of smaller budgets or estimates for operating costs and capital expenditures. Cost control, our sister subject, protects these decisions and holds expenditures within budget by constant (and early) monitoring and appraisal of the cost performance of those responsible for executing the projects.

One of the most striking examples of the potential effectiveness of cost engineering is the area of labor productivity. The impact of deteriorating productivity on a country's economy is well known. The adverse effect is significant, immediate, and disastrous. Yet, what is often not appreciated by all of the parties involved (management, labor, government, and the public) is the many factors (other than labor unrest) that have a significant effect on productivity and, more importantly, that are controllable. Perhaps through the vehicle of good cost-engineering practice in this area a better understanding can evolve, and those in a responsible position to do so, especially governments, can take the lead in developing programs that would result in significant improvements in this critical area.

The entire text is written on the premise that it is important for the reader to be given basic philosophy and ideas that he can apply to his individual needs. Consequently, there are few specific data included. We are advocating that the basic approaches outlined in the text be applied by the reader to develop his own data, estimating method, or cost control system. Most published data are based on historical experience of an individual company in a particular time frame under specific conditions. This text supplies the reader with the know-how to analyze these data, or this method, or this system, and determine what adjustments need to be made in order to make the data fit his or her specific needs.

Scheduling and cost engineering are closely related and intertwined. Consequently, scheduling is mentioned from time to time throughout the text. However, no details of schedule development, appraisal, or control are given, and the authors recommend the reader, if he or she is not already familiar with the subject, to become so by seeking out one of the many publications on this important subject. The same comment applies to the general area of Project Management. Much of what is presented in this text is based on experiences related to capital projects which have been executed utilizing present day Project Management techniques, which can also be found in publications on this equally important subject.

As mentioned above, the text includes a number of simple but very basic steps that if understood and followed can result in effective cost engineering. More important, through bitter personal experiences, the authors have learned that when these steps are *not* followed, management and the cost engineer are inviting financial disaster on a project. Consequently, we make no apologies for the constant repetition of these important basic points throughout the text.

Forrest D. Clark
A. B. Lorenzoni

CONTENTS

Part II: COST CONTROL

APPLIED
COST ENGINEERING

1

INTRODUCTION TO COST ENGINEERING

WHAT IS COST ENGINEERING?

The American Association of Cost Engineers (AACE) defines cost engineering as "that area of engineering practice where engineering judgement and experience are utilized in the application of scientific principles and techniques to the problems of cost estimation, cost control, and profitability." In this text, we shall be dealing with cost *estimating* and *cost control.* It should be noted that the above definition emphasizes that cost estimating and cost control are areas of *engineering* practice using *scientific* principles and techniques. The need for a strong engineering academic background will become evident as we discuss in the text the many techniques that must be applied to implement a successful cost-engineering program.

Because estimating or predicting cost and then controlling cost within the limits of the estimate is an essential part of any project and therefore any industry, cost engineering is not confined to any particular industry or group of industries and has application throughout the business world and, for that matter, the domestic world. Building projects without applying the principles discussed here can and often does result in financial chaos.

WHAT IS A PROJECT?

Throughout the text, reference will be made to building projects and project control. Projects can cover anything from a 2-day engineering effort to resolve a minor technical problem to the present-day "super projects," such as the $7 billion Alyeska pipeline in Alaska. Cost engineering has application regardless of industry; it also has application regardless of project size. In this text, we will

1

relate most of the cost-engineering principles to an average-size project in the process plants industry (i.e., a petroleum or chemicals plant in the range of $50 million to $100 million). However, where the techniques need to be altered or supplemented for smaller or larger projects this information will be included in the text. Again, we must emphasize that regardless of industry or size, the principles of cost engineering are the same and, once understood, can be applied to most situations.

PROJECT LIFE CYCLE

Using the $50 million to $100 million average-size process plant, Fig. 1 illustrates the typical life cycle of a project. Although the time spans will vary from project to project, the four phases shown will apply to all projects. This life cycle will be discussed in detail and referred to frequently in both the cost-estimating and cost-control sections of the text. Before discussing the cycle in detail, let us examine briefly the reasons why a corporation or company might decide to invest in and build a project.

WHY ARE PROJECTS BUILT?

There are a multitude of reasons why a corporation or company decides to build a project. The decision could be the result of a favorable analysis of the marketing situation that shows future increased product demands or the need for new and different products or the research department may develop new products with high sales potential, or new government or social requirements

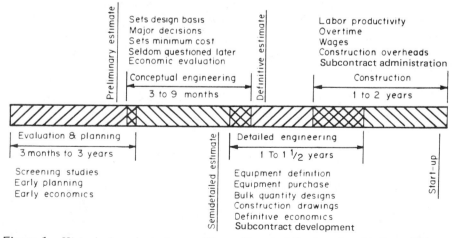

Figure 1   Historical project phases. (From Chemical Engineering, McGraw-Hill, July 1, 1975.)

may dictate the need for new facilities, especially in the area of environmental protection. Regardless of which of the above-mentioned reasons for building a project apply, the ultimate reason in virtually every case is economic (i.e., what is the maximum return that can be made for the money invested by the company?). In making this extremely important decision, management asks and needs the answers to many questions. How much product should I make? Where should the plant be built? How competitive will a new product be? If diversification is desirable, where and what?

To answer most of the questions raised requires economic studies. Economic studies cannot be made without investments, which in turn requires an estimate, and that in turn requires a cost engineer (CE). Not only is the CE needed during the early evaluation stage, but similar management-type decisions must be made throughout the life of a project. Couple this with the need for cost control during all phases and you have the need for cost-engineering services throughout a project life cycle.

THE ROLE OF CEs DURING THE EVALUATION PHASE

CEs play an important role in each of the four phases of a project shown in Fig. 1. Generally speaking, the earlier they are brought into the picture, the more effective they can be. Consequently, the first or evaluation stage can be the most critical. Unfortunately, as we will discuss later, this is the stage where often little or no cost-engineering assistance is sought by or made available to management.

The evaluation or planning phase covers the period from inception until a decision is made on a definite plan of execution (i.e., what will be built, where, and why). Because it is a period of screening, market evaluations, political and geographic considerations, and major high-level decisions (and indecisions!), the length of this phase is subject to far more variation than any of the other three phases. Although the chart shows an average of 3 months to 3 years, the authors are aware of projects having an evaluation period of only weeks and at least one example of a major project having been evaluated over a period of 20 years before a decision was made to proceed.

During the evaluation phase, the CE provides the cost estimates needed in the economic evaluations of a number of different areas. For example:

Different case studies reflecting alternate processes or alternate product yields
Different geographic locations
Evaluation of a preliminary sketch of a new process being developed, in most
    cases by a research or new-ventures department

In these evaluations, normally carried out by a planner or economist in the company, many factors have to be considered, and in most cases cost is the most important factor. It is during this stage that the highly critical project decisions are made, involving the largest amounts of money. Once the project is approved, future decisions can only involve a portion of the project; but at the early evaluation stage, a "make-or-break" decision is made on the total project value.

Unfortunately, this critical phase, when the most important decisions are made, is also the time when the least information is available to CEs and, consequently, unless a concerted effort is made, their estimates will tend to have a poor and often unreliable basis. Further, because those key people in the project are thoroughly engrossed in the selection process, CEs find it difficult to extract from them the necessary information to improve the estimate basis. In undisciplined organizations, CEs are often not consulted at all or only superficially during the evaluation stage, with the design or research developing their own estimates through prorations of costs from general published data, their own "gut" feel, or other similar undesirable means. Inevitably, CEs are brought into the picture, usually during a later stage of project development. Often they can then only play the role of "spoiler," uncovering the fact that a poor decision was made to proceed, the project is uneconomical, and it has to be scrapped at a hefty cancellation charge to the owner plus valuable lost time. It is essential to make CEs key persons in early project planning and to make a specific concerted effort to develop reliable data for the estimate basis.

Once a decision has been made to proceed (with the assistance and involvement of CEs), the project moves on to the next phase, that of establishing a firm design basis for the selected route.

THE ROLE OF CEs DURING THE BASIC-DESIGN PHASE

The basic-design phase is the stage in project development at which, for the first time, the project and management persons involved turn their attention to the specific basis for the selected route rather than the more generalized approach taken during the prior evaluation stage. During the evaluation or screening phase, the main emphasis was placed on the evaluation of delta costs of alternative approaches to select the most economical case. Little time or effort was spent establishing the basis for a total-cost picture. Having selected a route from these many alternatives investigated during the evaluation stage, the owner now needs to finalize a total investment based on the specifics of the selected case. This investment in many cases will be the basis for a request to higher management for project approval and therefore must be of higher quality than the earlier screening estimates. Let us examine what happens during the basic-design phase to help us understand why a better estimate can then be made. During this phase,

The process or basic design is finalized.

The physical layout of the facilities involved is given consideration.

The need for support facilities (utilities and off-plot items, such as storage, roads, and fire protection) is reviewed.

Timing (schedule) is given more serious thought.

The project execution strategy is also discussed (who does what, when, where).

You will note that we have used the terms "considered" and "discussed." Although during the basic-design phase the design basis will be finalized, many other areas that have a significant impact on costs will not yet be sufficiently developed to have a firm basis; for the purposes of making the estimate, CEs, with the assistance of the project people, will have to make certain assumptions. To understand how and why these assumptions are made, we need initially to discuss the first of a number of basic principles associated with the field of cost engineering (i.e., the definition of an estimate).

WHAT IS AN ESTIMATE?

An estimate is a prediction. In the case of a project, an estimate is a prediction of the manner in which that project will be executed. The estimate basis should reflect a step-by-step plan of how the project people feel or predict the job will be done. In the case of the design, they are putting down on paper a flow plan and equipment list, which predicts what the final design will look like after contractor engineering. The estimator takes the "hardware" list (augmented by hardware predicted by estimating methods during the conceptual-estimating phase, when the design is not as detailed) and translates it into a dollar value by applying unit costs to the quantities. In doing this estimators are predicting what will be paid for each item when it is purchased. They are also predicting when it will be purchased (escalation) and where it will be bought (world market source). They therefore need to develop a plan and schedule for that specific project.

The concept that an estimate is a prediction of the way something will be built will be used throughout the text in explaining many other cost-engineering concepts, such as the development of estimating methods and the application of cost control on a project.

IMPORTANCE OF DESIGN BASIS

During the basic-design stage, there are many potential variations in the basis that will be used for the estimate made during that stage. For example, the final site for the project may not yet be selected and it may differ physically and dollarwise from what is predicted in the estimate. Other variables that are not yet firm and consequently can vary widely are the project schedule, the availability and cost of labor, the source of materials, the final detailed design (vs. the conceptual basis design), and the amount of taxes and duty to be paid.

However, the single largest potential for variation, the design basis, is normally finalized at this stage. We are going to build one plant of a particular size, not two or three smaller sizes at different locations. The plant design will use a particular process, the selected one, and not any of the multitude of alternatives that were examined during the evaluation phase. Experience has shown that the variations in the design basis have a far greater impact on costs and schedule than the variations in other areas. Consequently, as we shall see later in this chapter, the fact that the design basis has been agreed to by those who have the power of change is extremely significant and has a dramatic effect on overall estimate accuracy.

DURING BASIC-DESIGN PHASE, CEs ARE JACKS-OF-ALL-TRADES

Because most areas, aside from the design basis, are not yet finalized, CEs and/or others involved in the project must do a considerable amount of predicting. Also, as was the case during the evaluation stage, the project people are engrossed in developing the design basis and have little time to devote to developing for CEs the details of the missing pieces in the other areas. Consequently, estimators are often left with the task of developing a "conceptual" estimate, wherein most of the predictions of what will be done are made by estimators or by their estimating methods or tools. As a result, CEs making the estimate act as "jacks-of-all-trades." For example, to predict the final hardware that will result from the basic design, CEs need to visualize what the final detailed design will look like (i.e., they must be a designer of sorts). They need to develop a time schedule for the project and therefore have to act as schedulers. They need to gather and analyze historical cost data and act as data analysts. They must establish costs for different geographic locations and countries and predict escalation rates, in effect becoming mini-world-economists. They must predict contract strategy: what kind of contract, to how many contractors, under what conditions, and act as contracts engineers. But, most important of all, to do all of the above they need data from and the cooperation of many people involved in the project. Therefore, above all, they need to be diplomats!

CEs' ROLE IN CONTRACTOR SELECTION

The next phase of project development after the basic design is completed is detailed engineering, which normally begins with the selection of a contractor by the owner or client. This phase is a crucial one for owners because, for the first time, they will be giving some of the controls on the job to another party (i.e., the contractor), so they are anxious to ensure that the selected contractor is capable of exercising good job control. Consequently, owners thoroughly review the qualifications of the various contractors who have been requested to bid on the job. An area that should, and often does, receive an in-depth review is the contractor's proposals for exercising cost control on the project. This will be discussed in more detail in the cost-control section of the text, but it should be noted now that this is an area where CEs are involved in reviewing the cost-control proposals and making a recommendation to management as part of the contractor selection process.

There are a number of other areas where CEs will be involved during contractor selection. If the bid is a fixed price or "lump sum" bid, owners need some guidance as to whether the bids received are reasonable (i.e., is the economic climate suitable for fixed-price bids or is there so little competition that the bids are not competitive and hence are too high?). To determine this, owners will have CEs prepare a bid-check estimate. This is a curve or semidetailed estimate, prepared over a short period solely to check the overall level of the contractor bids. We will discuss in Chap. 3 and 4 what these estimates are and how they are prepared.

Contractors may submit alternate proposals that may or may not be of significant benefit to owners. These alternatives have to be evaluated and this usually requires estimates that are prepared by CEs. If the contract is to be of the reimbursable type, CEs also assist the bid evaluation team in analyzing the commercial terms submitted by the contractors (i.e., the hourly costs of contractor engineering and field supervisions, the overhead costs, fee costs, etc.).

From the above, it can readily be concluded that CEs play a crucial role in the selection by owners of contractors for the project. As we will see later, this is only the beginning of what is to become a full-time involvement on the part of cost engineering during the more detailed project execution phase (a term used to describe the phase after contract award).

THE ROLE OF CEs DURING DETAILED ENGINEERING

After contract award, CEs embark on dual roles. We still need an owner's CE on the project (they are needed by the owner throughout the project to protect and monitor the economic decisions that were made much earlier by the owner), but now contractors require CEs to assist their project people to exercise cost control on the project.

During the detailed-engineering phase, contractors' CEs will prepare the definitive control estimate (if not already done by the owner, as is sometimes the case), and the estimates for change orders (which, unfortunately, occur on all jobs), and they will fill the key role of cost-control engineer for the contractor's project management team, a role that will be discussed in the Cost Control section of the text.

In the meantime, the owners' CEs will have the responsibility of reviewing and commenting on contractors' detailed control estimates, reviewing their estimates for extra costs and change orders and, most important, on a reimbursible project, appraising and monitoring the contractors' efforts in the area of cost control.

Despite the fact that the role of each of the cost-control engineers is considerably different, with that of the contractors' CEs detailed while the owners' CEs working in broader terms, they both are working toward a common goal (i.e., to complete the project at the lowest possible cost consistent with the project's objectives and, hopefully, within the project budget).

ROLE OF CEs DURING CONSTRUCTION

The last phase of project development is construction, when the hardware that was designed and procured during the detailed-engineering phase is installed by contractors in the field. Here again we have both the owners' and the contractors' CEs playing important roles. The owners' CEs continue to appraise the contractors' cost control efforts, monitor changes, and review extras. In addition to the areas of responsibility they had in detailed engineering, the contractors' CEs now become more deeply involved in providing input to the contractors' planning and scheduling team. Their forecasts of labor manhour requirements are an essential ingredient for the project schedule and have an impact on the prediction of both final costs and final completion date. This important role is discussed in detail in Chap. 20. In fact, the roles of CEs that we have discussed, from the early evaluation phase through construction, will be discussed in detail later in the text. However, before proceeding too far in this direction, it would be wise to list and briefly define some of the terms we will be using that have wide application in the cost-engineering world but may be unfamiliar or confusing to the reader.

SOME BASIC COST-ENGINEERING TERMS

The basic term "cost engineer" means one thing to an owner and often something else to a contractor. An owner's definition of a CE is an engineer who can and most likely will be involved in all aspects of cost engineering (i.e., cost estimating, cost control, and profitability). This is the definition we will be using throughout this text and is the one that is gradually being applied and accepted throughout the industry. On the other hand, most contractors still use the term "cost engineer" only to describe their cost-control engineer and refer to the engineer developing the cost estimates as an estimator. The reasoning behind this differentiation between owner and contractor will become more understandable after discussion later in the text on the manner in which owners and contractors set up their cost organizations.

Another basic area that is subject to different terms and definitions is types of estimates. As was explained earlier, there are three basic stages or phases to project development (if we combine the execution phases of detailed design and construction into one phase), and for each of these, estimates are made. Some of the terms used to describe these estimates are as follows:

Planning/evaluation stage
>*Screening*
>Preliminary
>Quickie
>Order-of-magnitude/guesstimates
>Rough, gross, scope, etc.

Basic design stage
>*Preliminary*
>Budget
>Semidetailed

Detailed engineering construction
>*Definitive*
>Appropriation
>Lump sum
>Detailed

Each of these will be explained in more detail in Chap. 3. Most other terms considered peculiar to the cost-engineering industry will be explained when first introduced in the text.

SUMMARY

Cost engineering as a discipline covers the areas of cost estimation, cost control, and profitability. In this text, we will be concentrating on cost estimating and cost control.

CEs should be involved in all phases of a project life cycle, from initial planning and evaluation to the start-up of the plant after construction.

An estimate is a prediction of the manner in which a project will be executed.

The single largest potential for variation in an early estimate's basis is the design basis. Finalizing the design basis, consequently, has a dramatic effect on estimate accuracy.

There are owner or client CEs and there are contractor CEs, with somewhat different roles to play but the same end purpose: completing a project for the lowest possible cost consistent with the overall project objectives.

Part I

COST ESTIMATING

2

ESTIMATE ACCURACY

WHAT IS A GOOD (vs. A BAD) ESTIMATE?

Before discussing in detail the subject of estimate accuracy, it would be well to establish what is generally considered to be a good, accurate estimate. Unfortunately, probably because human nature lends itself more to remembering bad experiences than good ones when it comes to estimates, there are more standards for bad estimates than there are for good ones. Generally, the standards for a bad estimate are the following:

Significant overrun on the original estimated value
Inconsistent results (large overruns and underruns)
Few details, especially where needed
Poorly documented
Unreliable for funds allocation
Unreliable for job control

There are a number of reasons given for a bad estimate, many of them misconceptions. Some of the more common misconceptions are the following:

Poor (unqualified) estimator
Poor estimate data
Poor estimate methods
Estimator unfamiliar with "real" life

In short, the estimator is to blame.

A good portion of Part I of this volume is devoted to dispelling these misconceptions and outlining the ingredients for a good, accurate estimate.

These ingredients, which estimators must have if they are to have any measure of success, are summarized as follows:

A sound finalized basis
A realistic execution plan
Proper timing
Good estimating methods and data
Neatly documented, balanced detail
A good, experienced cook (estimator)

Estimators play an important role in developing a good estimate, and we have discussed and will continue to discuss what qualifications are needed to enable them to take advantage of the ingredients listed above.

ESTIMATE ACCURACY vs. TIME

In Chap. 1 we learned that an estimate is a prediction of the manner in which a project will be executed, a very important concept in cost engineering. In this chapter we will discuss a second basic and important concept of cost engineering: the relationship between estimate accuracy and the time in the development of the project when that estimate is made, an essential consideration in developing the good estimate discussed earlier.

Figure 1 is a graphic illustration of this relationship. We have plotted accuracy on the ordinate, expressed as the probability that we will have an overrun in the estimate. On the abscissa we have shown a time scale, splitting the time of the development of a project into the three phases discussed in Chap. 1, [i.e., planning, basic design, and execution (contractor-detailed engineering and construction, the actual design and erection of the plant)]. Note that the curve expressing the relationship between the two variables indicates almost 100% accuracy for an estimate made near the completion of a project, an expected result. There are two important conclusions to be drawn from this curve:

1. The lengthy and vague time of planning is a period of a high degree of variability. Although not shown here the scatter of the actual data used to plot the curve in this period is wide and erratic. Consequently, not only is the probability of overrun great but the degree or magnitude of the overrun is difficult to predict.
2. There is a dramatic change in slope at the moment the design basis is selected and finalized, and a significant increase in accuracy occurs during the short period of time when the basic design is put together.

INGREDIENTS FOR AN ACCURATE ESTIMATE

Why does the accuracy curve have the shape that it does? To answer this, we must return to our first principle, the definition of an estimate: a prediction of

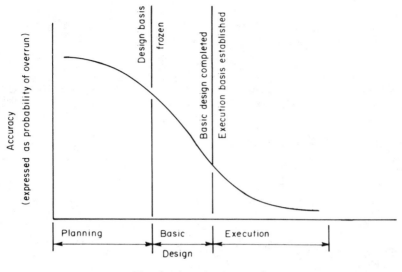

Figure 1   Estimate accuracy vs. time.

the way a job will be built. Assuming we have capable CEs doing the estimate, two factors affect the accuracy of the estimate: the reliability of the basis for their prediction and the reliability of the estimating methods and tools they use. Methods and tools are normally not the main culprit, at least not during the early stages of a project when estimate accuracy is poorest. They are important, however, and we have devoted Chap. 4 solely to this subject. The most critical item affecting estimate accuracy is the estimate basis.

During the planning phase, management, the planners, and the designers are vascillating between building a 10,000 bbl/SD (barrels per stream-day of product) and a 20,000 bbl/SD size plant, or building it in Germany vs. England, with perhaps no specific site and therefore site conditions in mind. Even the time when the plant will be built or what specific process will be used have not been determined. With these kinds of variables, it is not difficult to understand why an estimate prepared during the planning or evaluation stage has a poor basis and hence, poor accuracy. Nor is it difficult to appreciate that, even with the best estimating methods and tools and the cleverest CEs, it is impossible to make an accurate estimate if, several months later, the designer or management changes the capacity from 10,000 to 20,000. Consequently, as we mentioned in Chap. 1, the effort in the early phases needs to be directed toward establishing a firmer basis rather than concentrating on using more detailed estimating methods.

Figure 2 illustrates the same principles we have discussed but in a different and more detailed manner. The chart again shows that, as firmer basis information becomes available to CEs, the estimate becomes more accurate. In this figure, XYZ is equivalent to the planning or evaluation stage, F to the basic-design phase, and the remainder to execution. You will note the sharp increase in accuracy that occurs between XYZ and F, similar to the change in Fig. 1.

WHY IS KNOWING ACCURACY CRITICAL?

Knowing the accuracy of an estimate is critical to the management and the project people involved, including CEs. The reasons for this are many, and the most important are as follows:

Sensitivity to economics
Cash flow and budgeting
Determines the estimating method and the approach to be used
Feedback (improves method development)
Develops confidence/management support

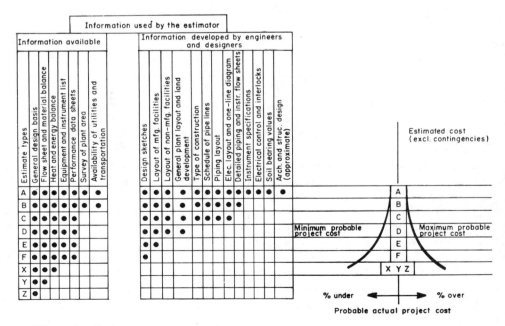

Figure 2   Estimate accuracy vs. time.

If we can develop an approach that allows us to determine in advance what kind of accuracy we can expect from a specific estimate made at a certain time in project development, we can then "test" the sensitivity of the investment level predicted by the estimate to potential variation and determine the impact on the calculated economics for the project. For example, if it is known (from historical, statistical analysis of past projects) that an estimate prepared during the planning phase has a variability range of ± 30%, the calculated economics for the base number (considered the most probable cost with a 50/50 chance of overrun or underrun) should be tested for a potential 30% overrun in investment. This is not an unlikely possibility, since our curve shows statistically that the probability is high. The curve in Fig. 1 represents the probability of overrun by a certain percentage, in many cases 10%. This 10% is for many companies the normal allowed overrun before management is forced to reappropriate additional funds. For example, the probability of overrun by more than 10% during the planning stage may be 3 in 4 but during the execution stage only 1 in 10. This says one should never risk an appropriation or major decision based on an early estimate. Unfortunately, this message is not always clear to top management. If the sensitivity check shows that the project cannot stand the lower range of return on investment, then management may reconsider, and cancel or change the direction of the project or, more likely, provide only a limited appropriation, sufficient to allow additional engineering and development of the project to provide a better design and execution basis and hence, a higher quality estimate.

Knowing the accuracy of an estimate can also assist management and others in predicting the cash flow for the life of the project. If the estimate is an early one, their predictions must be tempered by the inaccuracy of the estimate, as was done for the economics. This information is especially important in multimillion (and billion) dollar projects.

Knowing the limits of the accuracy of the estimate they are preparing is valuable information for CEs. When they are preparing an estimate early in the project life, they will recognize the degrees of variability that exist in the basis given to them and will use an appropriate estimating approach or method to prepare the estimate (i.e., a curve or "quickie" estimate will normally be made instead of a more detailed approach). This will be done regardless of the amount of detail supplied them. They realize that it is pointless to use the detailed basic information to make a more detailed estimate if the basis is subject to such a high degree of variability. A knowledge of when accurate estimates can and cannot be made can also be useful to CEs in developing the proper estimating methods and tools and in improving these methods. This subject is discussed in detail in Chap. 4.

Perhaps the most important reason that knowing the accuracy of estimates is so critical is the impact this knowledge has on company management and company policy and philosophy with regard to cost engineering. Once management and company personnel have been educated with the basic principle that an accurate estimate relies on an accurate basis and this in turn is a function of the time in the life of the project when the estimate is made, a remarkable change can and often does occur. Management now knows what to expect of an estimate when it is presented to them for review. No longer do they view an overrun in an early estimate as solely the result of a poor estimate done on the "back of an envelope." They will now put pressure on those responsible for setting the basis to insure they have given as much serious consideration to the basis as can be expected this early in the project life. More important, management will develop an understanding and appreciation of the problems associated with making estimates, particularly in the early stages of a project. They will not as a rule give full approval to a project based on an early estimate but will only give approval to proceed to an additional stage to get a better basis and estimate (i.e., go at least into the design-basis stage, if possible). If management, for other more pressing reasons, finds it necessary to proceed at an early stage, they now will be able to do so recognizing and analyzing the risks involved beforehand.

Because an informed management understands the many problems faced by CEs, they are in a better position to evaluate the support, both financial and moral, that must be given to the cost-engineering area within the company. As we will see in our discussions on estimating methods and cost control, this is of the utmost importance in developing a successful cost-engineering program in the company.

A very useful and effective technique for insuring that the best design and execution basis is available to the estimator and to communicate the importance of basis to estimate accuracy is to establish an estimate classification system within the company. In this system, all estimates would be identified or classified according to the reliability and the extent of the development of the design and execution basis information used in preparing the estimate. The quality of the estimating methods and data used would also be a factor in determining the estimate class. Since, in most cases, the quality of the basis, methods, and data used improves as the project is developed, it follows that the estimate class will also be related to the time in the development of the project when the estimate is made and, hence, is an indication of its accuracy and reliability, as explained earlier.

To illustrate this concept, Fig. 3 is a simple example of an estimate classification for the construction of a house. A class "D" estimate would be a early, rough order-of-magnitude estimate, where little known basis information is available and the majority of the parameters that determine the cost of the house will have to be assumed or implied within any estimating method used. As time progresses and more information becomes available, a better class estimate can be prepared, until eventually all the critical definition is available and an accurate and reliable Class "A" estimate can be prepared. Looking back at Fig. 2,

House construction Project definition	Estimate class Minimum information required			
	D	C	B	A
Size (sq. meters)	•	•	•	•
Location	○	•	•	•
No. bedrooms	X	○	•	•
No. baths	X	X	○	•
Exterior materials	X	○	•	•
Type contracting	X	X	○	•
Construction details	X	X	X	○

Figure 3 Simple example of an estimate classification for the construction of a house.

the same principles could be applied and the "estimate types" could become the estimate class for a classification system. The blank spaces would be filled in, with an indication as to whether, at that stage, the missing definitive information is available in a preliminary form or must be assumed or extrapolated from previous experience or historical data.

These are simple examples. Most estimate classification systems would be more comprehensive. For each class, every design and execution item which is investment-sensitive would be listed and described in some detail. For a given class, the items listed would have to be available, in the detail described, to estimators for their use in preparing the estimate and for that estimate to attain the particular class. If information is lacking, then the estimate is given a lower classification or, alternatively, the estimate is delayed until the missing information is developed.

An estimate classification system is a powerful tool and, once properly developed and effectively implemented, can have the following significant advantages:

1. Serves as a discipline within the organization by insuring that the available basis information is thought out and documented in an organized fashion, not only for estimators but also for project execution planning.

2. Clearly highlights at each stage of project development the missing basis information and additional amount of work required to prepare a given class estimate.

3. Provides an excellent means of communicating estimate quality within the organization. Estimates will be identified and discussed in terms of the quality of their basis and their accuracy rather than described by their end use. For example, the term "budget" estimate only means that the estimate will probably be used to allocate or appropriate funds. The estimate could be a good Class "A" estimate or a poor Class "D" estimate. In most cases, without the discipline of an estimate class, it will likely be a mixed bag and it is therefore often misleading to use terms like "budget" to describe an estimate.

4. Allows management to review estimates and make rational economic decisions with the full knowledge of the quality of the estimate basis and, hence, the estimate accuracy.

5. Provides a basis for analysis of the actual vs. estimated results, since, by using an estimate classification system, there will be a degree of consistency in the preparation of all estimates. The net result is that, eventually, the accuracy of each class of estimate can be established and the contingency levels determined.

6. As mentioned earlier, knowing the accuracy allows testing of the sensitivity of the investment level, is important to cash flow and budgeting, helps determine which estimating approach to be used, improves feedback and hence methods development and importantly, develops confidence in and management support for the cost-engineering area within a company.

All of the above reasons are of sufficient importance to warrant consideration of the establishment of a formal estimate classification system within a company organization.

ACCURACY IS A FUNCTION OF INDIVIDUAL EXPERIENCE

As mentioned earlier, there are three factors that impact on the accuracy of any estimate:

1. Time in the life cycle
2. The CE
3. The methods and tools available

All three are factors that can vary considerably from company to company. For instance, the number of estimates made, the involvement of the CE, and the emphasis on establishing a reliable basis vs. other project priorities can change the shape of the curve in Fig. 1. Also, the qualifications of the CE and the quality and reliability of the estimating methods the CE will be using will, to a large degree, be determined by the emphasis and importance that the individual company places on cost engineering. These, in turn, will have an effect on the shape of the accuracy curve. Consequently, we must conclude that each company should develop its own accuracy data and curve. However, the basic principles—that the earlier in project life an estimate is made, the less accurate it will likely be; and that a finalized design basis improves accuracy significantly— would apply to all companies.

SUMMARY

The following is a summary of the more important principles discussed in this chapter:

Estimate accuracy is a function of the time the estimate is prepared in the project life cycle.

Estimate accuracy is affected more by variations in basis than in estimating methods or the estimator.

The earlier in project life an estimate is prepared, the less accurate the basis (and hence the estimate) is likely to be.

Estimate accuracy increases dramatically after the design basis is finalized.

Estimate accuracy is a function of an individual company's experience and its approach to project execution.

Knowing estimate accuracy is extremely important to company management, the project management team, and the CE.

An estimate classification system is an excellent communication and discipline tool and can result in imporved estimate quality and accuracy.

3

ESTIMATE TYPES

THREE MAJOR ESTIMATE CATEGORIES

In Chap. 1 we discussed briefly the three basic types or categories of estimates relating to the three major phases of the project life cycle. These three estimate categories were for estimates made during the following stages:

1. Planning/evaluation stage—*screening* estimates
2. Basic-design stage—*budget* estimates
3. Detailed engineering/construction stage—*definitive* estimates

Chapter 1 also listed a number of alternate names that are given to each of the above categories but, in general, the terms screening, budget, and definitive are the more commonly accepted terminology used in the industry. Let us now examine each of these categories individually and discuss what approaches and techniques might be used in preparing the estimate.

SCREENING ESTIMATES

The screening estimate is the earliest of estimates and, as the name implies, is made to enable the user to decide which way (and whether) to go on with a project. As such estimators are often interested either in a rough, order-of-magnitude total cost (if it is a single estimate and the "screening" is a question of go or no-go for the particular project) or, as is more often the case, they may be concerned primarily with the rough delta cost between two or more alternative cases. Management usually requires this information to select the proper route for the next step in developing the project. Regardless, in either case, the emphasis is not on detailed accuracy but rather on a reasonable cost level or delta cost of sufficient accuracy to insure that the results are meaningful and, above all, not misleading. To accomplish this, three different approaches can be used to prepare a screening estimate:

1. Gross (overall) proration
2. Curves
3. Rough semidetailed

The most commonly used approach is the gross proration. Curves and the semidetailed approach, although often used, are more commonly associated with budget estimates. Consequently, we will use the gross proration as an example of the typical approach for preparing screening estimates. Figure 1 illustrates how *not* to prepare a gross proration while Fig. 2 serves to correct the errors of Fig. 1 and outlines how a gross proration estimate should be done. We have shown both because, unfortunately, Fig. 1 is more representative of what is done than Fig. 2.

In our example, we assumed a typical situation. An owner's marketing arm has decided that there are or will be additional outlets for the company's products in a given area and, initially, management is interested in getting a feel for the amount of money involved and the rough economics of building a new pipestill in their existing refinery on the outskirts of Philadelphia. (A pipestill is a first-stage, key process plant in a refinery that distills the crude oil into heating oil, gasoline, etc.) The estimator is given the problem of determining the investment required to build the new plant. Theoretically all the estimator knows is the plant size, type, and location (which, from Fig. 2 of Chap. 2, is sufficient to make a screening estimate). The estimator's approach (in Fig. 1, the "bad" example) is simple. The company had built a similar plant near the city of Rotterdam, the Netherlands, in 1979 and, although it was larger, the estimator was able to extract the final cost of the plant from historical cost records and prorate that cost based on the capacities of the two plants in the following manner to arrive at the answer. Now let us examine the way the estimator

---

Problem: estimate cost of 100,000 bbl/day pipestill (P/S) at Philadelphia

Solution: prorate from 150,000 bbl/day P/S at Rotterdam

150,000 bbl/day P/S—Rotterdam—final cost—1979 = \$50 million

Cost—100,000 bbl/day at Philadelphia $= \dfrac{(100)}{(150)}^{0.55} \times 50 = \$40$ million

---

Figure 1   Gross proration, a bad example.

Problem: estimate cost of 100,000 bbl/day P/S at Philadelphia

Solution: prorate from 150,000 bbl/day P/S at Rotterdam

150.000 bbl/day P/S—Rotterdam—final cost—1979	= $50 million
Deduct piling, owner costs, tankage	= 10 million
	40 million
Escalate to 1986 = +150%	= 100 million
Rotterdam to Philadelphia = × 1.30	= 130 million
150,000 bbl/day to 100,000 bbl/day $= \dfrac{(100)}{(150)}^{0.55} \times 130 =$	104 million
Add pollution requirements	= + 11 million
Deduct credit for competitive climate	= − 6 million
Total cost =	109 million
Call	110 million

Figure 2   Gross proration, a good example.

should have approached the problem (Fig. 2). The difference here is that the estimator is taking a few minutes to think out the way the job will be executed, (i.e., she is trying to predict what will happen). In doing this, she is coming up with a much improved, more accurate, and realistic estimate. Let us look at what she has done.

First of all, she immediately recognized that, although the two plants were similar, there were some significant differences. The Rotterdam plant had some extra items that were not required for her plant in Philadelphia: piling, tankage, and certain owner costs, all significant cost items. After deducting these, she then adjusted for the fact that her new plant would not be completed until 7 years after the Rotterdam plant. She did this by adding her estimate of the average escalation over that period of time. The next important step the estimator took was to recognize that the locations of the two plants are in two different parts of the world, with different economic conditions and hence a difference in the cost of the plant. In this case, she applied a "location factor" or multiplier of 1.30 to being the now 1986 Rotterdam plant to Philadelphia.

However, she still had too large a plant—150,000 bbl/SD vs. the 100,000 bbl/SD required. To correct for this size difference, it was necessary to prorate the cost based on the relative capacities of the two plants, the only step that was taken in Fig. 1. In both cases, the estimator used a proration factor or slope of

0.55, a value she calculated or derived from analysis of the behavior pattern of the cost of similar plants as they change in size. Proration factors or slopes are extremely important in early estimates and are used not only for making gross prorations but also in the development and preparation of curve, factored, and semidetailed estimates.

The proration factor to be used in the case of a plant is the slope of the curve that has been drawn to reflect the change in cost of that plant if it is made smaller or larger. The curve is usually drawn based on data points of the cost of completed plants or detailed estimates for new plants. The slope of the curve for a plant is the summary or net effect of a multitude of slopes of each of the plant's components. For example, some components that make up a plant have a flat slope, such as compressors or instruments. These items do not change significantly in cost as they or the plant increase in size and can have slopes as low as 0.3 or 0.4. On the other hand, items such as furnaces or piping can have slopes as high as 0.7 or 0.8. By analyzing and weighing the individual components, the estimator can develop the overall plant slope. If the plant is a common one, she can probably find the slope in published literature. In any event, this information is essential to the preparation of early estimates.

Returning to our good example of a gross proration, our estimator is still thinking through, or predicting, what will happen if the plant is built. She knows, for instance, that directionally there are more stringent pollution requirements in the Philadelphia area, which will add roughly 10% to the total cost. On the other hand, she also knows that economic activity is much lower and hence the climate is more competitive in the United States in 1986 than it was in Rotterdam in 1979 and this is worth about a 5% reduction in total cost. Although, in this theoretical example, these two items are offsetting and the reader may be questioning the value of carrying out these steps, they are essential. They could easily have been additive, making a plus (or minus) difference of 15%. Also they require no special effort and little time to calculate.

Table 1, page 27, summarizes the normal conditions under which a screening estimate using the proration approach may be used. It also summarizes the advantages and disadvantages of this method. A fuller appreciation of these will be developed as we examine the techniques and approaches used for budget and definitive estimates.

BUDGET ESTIMATES

Budget estimates are normally prepared after the owner has completed his planning work, has screened out his options (via economics, utilizing the screening estimate as a basis), and is now in a position where normally he will require management approval to proceed in developing the project further. To do this, he needs to give management some idea of the cost of the specific project he

Table 1   Prorations

Need historical data on similar plant/process
Must be near duplicate
Should be reasonably close in size
Proration slope critical, can be analyzed
Must adjust for off-sites/utilities
Must adjust for project execution differences (*prediction*)
Must escalate, adjust for location

Advantages	Disadvantages
Quick	Degree variation ($\pm$ 40%)
All-inclusive	Easily misused
	False sense of security

has in mind so that management can allocate or budget money in the future for this purpose. Hence, most estimates prepared during this time are called "budget estimates." As explained in Chap. 1, these estimates, done during the conceptual- or basic-design stage, are based on a finalized design (i.e., a selected route—no more alternatives or screening cases) but still lack a considerable amount of the detail and specifics. However, as discussed in Chap. 1, finalizing of the design basis contributed significantly to the accuracy of the estimate. In a budget estimate, we need some accuracy in the total estimate. With screening estimates, we were primarily interested in delta costs but now management wants to allocate or budget the total cost for the specific project proposed. To take advantage of this increased accuracy afforded by the finalized basis, there are three commonly used estimating methods for preparing a budget estimate:

1.   Curve
2.   Semidetailed
3.   Factored

All three methods are used with equal frequency throughout the industry and, further, because there are probably far more budget estimates made by both owners and contractors than screening or definitive estimates, we will examine the approach to be used for all three methods.

Figure 3 illustrates how a curve estimate would be made for our same 100,000 bbl/SD pipestill project at Philadelphia, discussed earlier as our screening estimate example. The curve has been plotted from past historical data (final costs and/or definitive estimates) for similar pipestills. The ordinate is total erected cost (TEC) while the abscissa is some variable or parameter in the design of the pipestill that impacts the most on the cost of the unit; in the case of a pipestill, the feed to the unit and the amount of overhead product. The

Problem: estimate cost of 100,000 bbl/day P/S at Philadelphia

Solution: use P/S investment curve for on-sites, % for off-sites/utilities

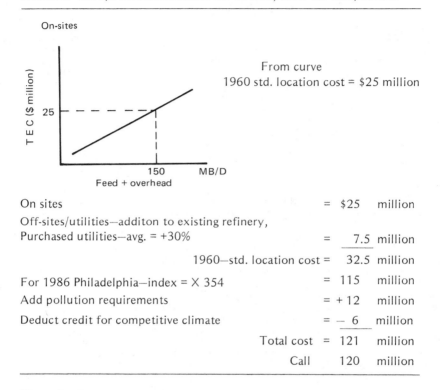

On-sites

From curve
1960 std. location cost = $25 million

On sites		=	$25	million
Off-sites/utilities—additon to existing refinery,				
Purchased utilities—avg. = +30%		=	7.5	million
	1960—std. location cost =		32.5	million
For 1986 Philadelphia—index = X 354		=	115	million
Add pollution requirements		=	+ 12	million
Deduct credit for competitive climate		=	− 6	million
	Total cost	=	121	million
	Call		120	million

Figure 3   Example of curve estimate.

curve is only for the on-site portion of the plant (i.e., the process unit facilites, usually within a well-defined boundary or battery limits, but excluding off-site auxiliary or support equipment such as storage tanks, product-loading facilities, long pipelines, steam or power generation, etc.). Historical data for on-site facilities lend themselves to the averaging concept that is inherent in investment curves, whereas off-site facilities are more customized and require a different approach, such as the simple "percentage of on-sites" used in the example in Fig. 3. You will note that the costs on the curve reflect a base year (1960) and a standard location. To arrive at the cost for a specific time (1986) and a specific location (Philadelphia), an index is used. As we will discuss later in Chap. 7, the use of base years and a standard location for the base of a company's estimating methods is common and essential where many locations must be considered.

Our example is basic and simple and can, with a small amount of additional effort, become more sophisticated and hence more useful and accurate. Figure 4 illustrates how our pipestill on-site curve can be split into three separate curves to give both more detail and, providing that the additional parameters are known, a more accurate estimate. Off-sites also can be done in more detail and accuracy, as shown in Fig. 5. The parameters required for these curves (i.e., total pounds per hour of steam to be generated, total barrels of tank storage, and the barrels per day of product) are usually known with a reasonable degree of accuracy at the time the budget estimate is prepared. Except for the first overall TEC curve of Fig. 3, all of the other curves give the estimate of the M + L or direct material and labor costs. There are other costs that must be added to arrive at a TEC. These costs are normally referred to as "indirect costs" and consist of such items as engineering, overheads associated with the field installation, fees, and contingency. We will discuss each of these in more detail later in the text. For our curve estimate example using the individual M + L curves, these indirect costs can be added as a percentage of the M + L costs, either individually, if historical percentages are available, or as one overall percentage addition.

Curve Estimates

Table 2 summarizes the significant points regarding curve estimates. It should be emphasized that curve estimates, like all other estimates, predict the execution of the project. Consequently, when the estimator picks a point on the curve for her project, that point in reality represents a certain amount and type of physical hardware, consistent with the average of what is in the design of the plants that make up the curve. This concept becomes important in the cost

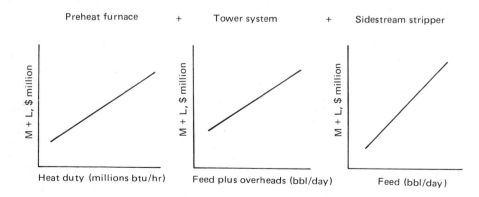

Figure 4   On-site curves can be made more sophisticated (i.e., for a pipestill).

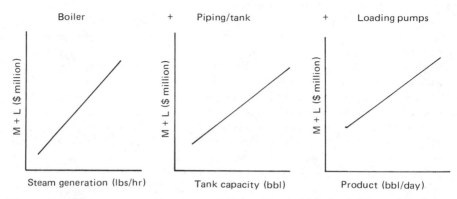

Figure 5 Off-sites curve estimate possibilities. Use: yield chart data and design utility requirements.

control approach to be used during conceptual design. As long as one has some idea of the details of the estimate prediction, one can then compare what is actually happening against what was predicted (the estimate). This is an essential element of cost control, as we will discover in later chapters.

Factored Estimates

Factor estimating relies on the principle that a ratio or factor exists between the cost of a particular equipment item and the associated nonequipment items that must be added to the project to complete the installation of that item. Hence, the term "ratio estimating" is often applied to factor estimating. For example, when, say, a drum is added to an estimate, the nonequipment items

Table 2   Curve estimates

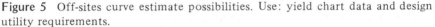

Need significant historical data	
Really is a hardware predictor	
Can be sophisticated	
Can be accurate	

Advantages	Disadvantages
Quick	Some variation ($\pm 20\%$)
Reasonable accuracy	No details available
All-inclusive	More on-site-oriented

that are also added as a result of including the drum in the project are the following:

A foundation for the drum to sit on
Excavation for the foundation
Instrumentation for the drum
Structural steel for the drum platforms
Piping to connect the drum to other process vessels
Supports for the piping
Paint and insulation for the drum and for the steel and piping

In addition, pro rata shares of other nonequipment items must be added, such as an incremental amount of paving, sewers, control house, and electric lighting. Whatever impact there is on the cost of material and labor created by the addition of the drum must be included in the estimate. This is done for each equipment item and, like a series of building blocks, each of the total M + L costs are added together to arrive at the overall total M + L for the project.

Figure 6 is a brief example of a factored estimate. Here we have extracted a small portion of the flow plan for our same 100,000 bbl/SD pipestill at Philadelphia. You will note that we do, in fact, have a flow plan, a significant improvement in basis information over our previous examples of gross proration and curve estimates. For each of the items on the flow plan, the estimator also has sufficient data in the form of equipment size, design pressure and temperature, and materials or construction. This will enable her to establish the cost of the individual equipment items, either using her own estimating methods or securing preliminary estimates or quotations from vendors. The estimator then takes the estimate for each equipment item and multiplies it by a factor (different for each equipment item) to arrive at the total M + L (including nonequipment items) associated with that item. By adding each of these "building blocks" together, she arrives at a total M + L of $78 million. To this she adds the indirect costs, estimated in the manner discussed earlier, to arrive at a TEC of $115 million (note that FLOH in Fig. 6 is a commonly used abbreviation for field labor overhead).

The example we have shown in Fig. 6 is the simplest of approaches and can be improved. More sophisticated factor estimates can and often are made by using one or more of the improvements listed in Table 3. In each of the items listed, the estimator is taking advantage of the data she has for her specific project. In most cases, the factored estimate is made at a stage in project development when at least equipment sizes and design conditions, and often other data such as piping sizes and instrumentation are known. Wherever possible, this information can and should be used in lieu of the average conditions inferred by the overall factors used in Fig. 6. The net result is a signifi-

Problem: estimate cost of 100,000 bbl/day P/S at Philadelphia

Flow plan

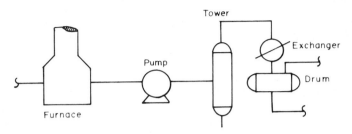

Solution: use factored estimate

		$ thousand
Furnace-detailed estimate (attached) = $650		
Total M + L = 650 × 2.2 =		1,430
Pumps-detailed estimate (attached) =	50	
Total M + L = 50 × 3.0 =		150
Tower-detailed estimate (attached) =	120	
Total M + L = 120 × 2.9 =		350
Etc. for all other equipment		
Total M + L =		78 million
Indirect costs [engr., FLOH, etc. (attached)] =		35 million
Total cost = $113 million		
Call = $115 million		

Figure 6   Factored estimate example.

cant improvement in estimate accuracy, since the estimator is reflecting the specific conditions applicable to her project.

Table 4 presents a brief summary of factored estimates. The reader's attention is drawn to the main disadvantage of a factored estimate (vs. a curve estimate, the normal alternative at this stage in project development). The curve estimate, because it is developed from final costs or detailed estimates of completed plants, is all-inclusive and very likely will not exclude any major items that will eventually show up in the detailed design of the specific plant estimated. The factored estimate, on the other hand, is, as we have discussed, a series of building blocks. If any of these blocks are omitted, the estimate will be low by that amount. Further, the interface between building blocks is not

Table 3 Factored Estimates: How Can We Do Better?

Vary factors as equipment size varies (use curve)
Reflect specific materials of construction
Develop factors for M and L
Adjust for differences in piping
Adjust for significant change in instruments
Adjust for known site differences
Develop factors for, and include, *all* equipment items

always clearly defined, and here again omissions may occur. Consequently, a factored estimate will have a tendency to be biased low, a condition common to most early estimates, but one to which the factored estimate is most vulnerable. Unfortunately, because this type of estimate takes more time and provides more detail, the user is often misled by a false sense of security (vs., say, a curve estimate, which is done in a fraction of the time and has virtually no detail). However, once the shortcomings are known and taken into consideration, the factored estimate can be the backbone of an organization's budget-estimating methods.

Semidetailed Estimates

Semidetailed estimates are similar to factored estimates, with one notable exception. Semidetailed estimates will attempt to predict the quantities and costs of the individual nonequipment or bulk items rather than lump these into factors or ratios. The bulk material quantities and costs are generated based on historical data, using physical size as a parameter (i.e., drum volume, pump horsepower, etc.) and relating this to the average bulk material quantities from

Table 4 Factored Estimates

Really ratio estimating
Need equipment sizing, at minimum
"Predicts" nonequipment items
Can be made sophisticated/accurate

Advantages	Disadvantages
Reflects specific design	Must account for all items
Less time than definitive	Tendency not to adjust
Some increase in accuracy ($\pm 15\%$)	Lacks detail for bulk materials

past designs. Because of the large amount of repetitive calculations, semidetailed estimates are often programmed for a computer and there are a number of commercial programs on the market. In most cases, these programs can (and should) be adjusted to reflect the individual company's way of doing business and its own needs. Modifications can also be made by the estimator to reflect specific bulk material data, if it is known and considered to be significant.

The most significant advantage that semidetailed estimates have over factored estimates is the provision of average bulk material quantities and costs. As will be seen in Part II, this information is essential for early and effective cost control.

DEFINITIVE ESTIMATES

The last type of estimate we will discuss, the definitive estimate, is the most accurate and the one that requires the greatest effort to prepare. Whereas the screening and budget estimates can be completed in a matter of hours or days by one estimator, definitive estimates require months to complete and, for larger projects can require an effort of thousands of technical manhours.

There are basically two kinds of definitive estimates. One uses the conceptual or averaging approach, is often referred to as an "in-house" estimate, and is used mainly by owner organizations. In this approach, estimating methods are developed from past historical data and an averaging concept is applied. It is more or less an extension of the factored estimating approach, except in greater detail.

The second type of definitive estimate is much more detailed in its approach and is used mainly by contractor organizations. Here, each item for the specific project is developed engineering-wise to sufficient detail to allow a customized estimate to be made of that item. Vendors' preliminary estimates or quotations for individual equipment items are widely used. Preliminary takeoffs are made of piping, etc., from early detailed-engineering drawings, and these takeoffs are priced up to form the estimate.

Whereas the conceptual approach requires a significant effort in the data-gathering and methods development areas beforehand, the detailed approach requires a large effort during the preparation of the estimate. The actual methods used in both types are discussed in greater detail in the next chapter. In this chapter we wish only to highlight the advantages and disadvantages of both approaches. The in-house approach has the advantages of requiring considerably less time and effort to prepare and, as we will learn in Part II, is more suitable for cost control. The disadvantages are that a major data-gathering and methods development effort is needed. Also, the end result is less accurate than the detailed approach (averaging vs. specific data). The detailed approach used by a contractor, on the other hand, is more accurate but has the disadvantages of being costly to prepare and often being available too late for early cost control on the project.

SUMMARY

There are three types of estimates:

Screening
Budget
Definitive

The approach used in making any of these estimates depends on the end use of the estimate and the need for an accurate overall total vs. the need for details (for control purposes). For example, a curve estimate could under certain condition be accurate enough for allocating funds (setting a budget) but will not provide the details for cost control. If no details are required, then additional effort should not be expended on a factored or definitive estimate to arrive at a slightly more accurate total nor should a detailed approach be used where the estimate basis is poor and uncertain. The approach used will also depend on the estimating tools available, the time available to prepare the estimate, the money

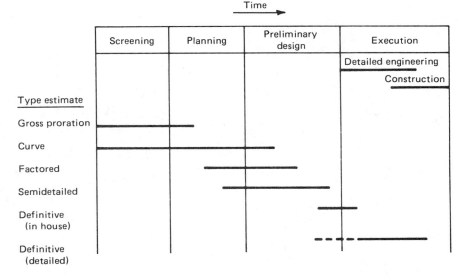

Figure 7   Types of estimates and when they are used in relation to the time in the life cycle of a project.

available for preparing the estimate, and the amount of previous historical cost data available.

Figure 7 summarizes the types of estimates and when they are used, in relation to the time in the life cycle of a project. Chapter 4 discusses in detail how and where a company can invest its effort to tailor methods to meet its individual needs.

4

ESTIMATING METHODS

WHAT IS AN ESTIMATING METHOD?

One of the key functions that is essential to the development of a good cost-engineering system within an organization is the establishment and use of sound estimating methods and data. An estimate is composed of essentially two ingredients, quantities and unit prices. The estimate of the quantities or "hardware" is accomplished by using an estimating method, whereas the estimate of the unit price to be applied to the quantities is determined through the gathering and analyzing of data. The latter item is the subject of Chap. 7.

Let us first describe what an estimating method is. We define it as a systematic and consistent approach to predicting or estimating the cost and schedule impact of overall job execution. For each item, whether it be equipment, bulk materials, labor, engineering, or any other item or category in a plant, a method is developed which in effect is a "pre-estimate" for that item, based on the average quantities required from past, good designs. Hence, most methods are based on an averaging concept. Most estimating methods are "in-house" methods (i.e., they allow the estimator to complete the estimate almost totally within his own department, with little or no help from other departments or from vendors and others outside the company).

However, there are other kinds of methods or approaches currently in use in the industry. One method that we do not recommend but, unfortunately, is commonly used is a system that is allowed to grow in a haphazard fashion. As new and experienced CEs are hired into the organization, they bring with them their background knowledge, often in the form of a collection of notes, copies of past estimates they have done, and published data. This type of information becomes the core of the company's estimating methods, especially when CEs become the chief estimators for the organization. Because this information is not centralized or formalized, it is referred to as "bottom-desk-drawer" data, the location of the data bank for the cost-engineering group. For many reasons, experience has shown that a more formal and systematic approach to the gathering of a data and the development of estimating methods is more than justified for most companies.

At the opposite end of the scale is the situation where a company sets up a methods group whose sole purpose is to develop and maintain in-house estimating methods. What would this group do? They would, for example, systematically examine each estimate category and develop a method of preestimating that category which could then be used by all of the estimators in the company. The method would reflect the manner in which that company executes a job and would, therefore, be based on past jobs done by that company. Published methods developed by others could be used, but only after modifying the method to tailor it to the company's particular experience and needs. This is essential if the final product, the estimate, is to be as we have defined it (i.e., a prediction of what will happen on the particular company's project).

WHY ARE ESTIMATING METHODS CRITICAL?

There is little doubt that the backbone of an estimate is the methods used in preparing it. After estimate basis, method is the most important element. In fact, it is good estimating methods that allows the estimator to take advantage of a good estimate basis and, hence, prepare a good estimate. There are many reasons (as listed below) that estimating methods are important and justify the effort to produce them:

Consistency
Accuracy
Time saving
Money saving
Availability
Confidence
Morale
Training

Let us take a brief look at each of these.

Since the methods developed would be used by all the company's CEs, all of their estimates would be consistent in format and approach and, as such, would make it much easier to analyze the results (actual vs. estimated). This is important in increasing the accuracy level of the estimates prepared by the company. As shortcomings are found, they can be corrected based on feedback, and this "fine tuning" of the method will result in further improved estimate accuracy. Since the method is essentially a preestimate based on historica data, estimators do not have to "reinvent the wheel" each time they estimate a concrete foundation or a structural support. It is done *once* by the methods group in a format that allows estimators to take advantage time-wise of the pre-estimating. The savings in time, in most cases, more than offsets the additional cost, in the long run, of the methods group. It also makes the estimate available much sooner, with all the advantages an early estimate provides to management and to job control, especially for planning and scheduling.

It is to management and the cost engineers that the greatest advantages accrue from good estimating methods. Once management understands the methods used, they and the CEs can more readily separate "the wheat from the chaff," (i.e., be in a much better position to objectively analyze an overrun in a project and determine whether the problem was in estimate basis or as a result of a poorly prepared estimate). The net effect is greater management acceptance and support for the cost-engineering function and an increase in the morale of the CEs.

Methods are also useful in training new, inexperienced CEs. They provide an immediate, consistent introduction for the new engineer, not only to the estimating approach but also to the company's way of doing business, since this is, in effect, what the methods are predicting.

One last advantage is not mentioned in our earlier listing. Clearly thought-out and documented methods lend themselves readily to computerizing, with all of the associated further advantages in timing, cost, and consistency. This subject is covered in more detail later in this chapter. Now that we understand what a method is and why it is of such importance to cost engineering, we will review how some methods are developed.

THE AVERAGING CONCEPT

The most common approach used to develop an estimating method is the averaging concept. This concept is best explained by example. Figure 1 gives a method for estimating the quantity of concrete required for pump foundations. The curve was developed by first determining what variable associated with the pump has the most significant impact on the size of its foundation. This was found to be the size of the pump, expressed most easily by the size of its driver in horsepower (hp). By examining a large number of past pump foundation designs on previous projects, a relationship and curve was developed. While investigating the data, it was discovered that a significant difference existed between piled and unpiled foundations and, as a result, a second curve was developed. If data points for either of these curves were found to be lacking from historical data, then the CEs developing the method requested that designs be prepared by civil designers for the missing items. Although all of this represents

Figure 1   Pump foundation (centrifugal).

some effort, it must be remembered that this is done only once. Once the curve is completed, it will be used over and over again for all future estimates. Another important point is that all the data for the curve came either from the company's past projects or from the company's own designers, thereby reflecting the average of the company's performance in the design of pump foundations. Any abnormal data would be excluded. All of this again is important if the estimate is to be an accurate prediction of actual performance.

Figure 2 is another example, this time for an equipment item, a shell and tube heat exchanger. Here the data are derived from past purchase orders, escalated and adjusted to a common base point. The most suitable variable is the square footage of the tube surface. Note that the curve is for a particular material of construction (carbon steel or C. S.), certain tube size and layout, and given temperature and pressure conditions. The CEs would develop an accompanying table with factors to adjust for differences in these key areas.

PIPING: A SPECIAL PROBLEM

Piping is the heart of any process unit and the estimating of the quantities of piping (i.e., pipe, valves, flanges, and fittings) is critical to arriving at an accurate estimate. Piping can equal 20 to 25% of total direct-material costs and 40 to 50%

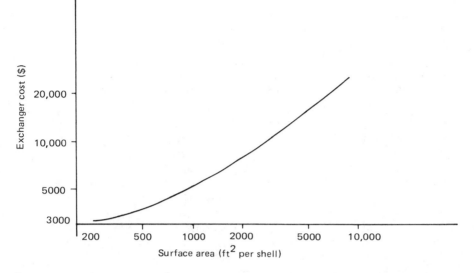

Figure 2   Shell and tube heat exchanger cost (TEMA-AES).

of total direct-labor costs. In addition, there is a direct relationship between piping quantities and the support for the piping, the foundations for the supports, and the paint and insulation for the piping. So the degree of accuracy in predicting the quantities of piping required has a direct and significant impact on the overall estimate total. For many reasons, however, the amount of piping required from project to project has a high degree of variability and consequently is difficult to estimate with consistency and accuracy. Layouts vary from plant to plant and even where they are similar, the design produced by one contractor can be significantly different that that of another, a key factor in cost control. The sheer quantities involved are overwhelming. There are thousands of piping items, whereas there may only be a dozen drums or two dozen pumps in a project. Developing a method that consistently and accurately predicts these quantities is not an easy task. As a result, control of many projects has been lost because of a poor or inadequate piping estimate. One method that has been developed and used with a large measure of success by a large owner company is the "unitized" piping estimating method. Because it is so successful and is representative of the results one can achieve through estimating method development, we will devote a portion of this chapter to explaining and discussing unitized piping.

Unitized Piping

To develop the unitized piping method, as was the case with our previous examples of averaging concept methods, a complete review was made of past piping designs made for the owner by various contractors on previous projects. From this review, standard piping patterns were developed for each piping service (i.e., for a pump suction line, a pump discharge line, a tower overhead line, etc.). The average quantities (e.g., feet of pipe, number of flanges, fittings, and valves) were established for each service. Figure 3 illustrates how a process flow plan can be broken down into typical piping patterns. Pattern types have been assigned to each line number on the flow plan. Figure 4 shows how quantities have been assigned to each typical piping service or pattern number. All of these quantities represent the average expected in the design, except for the valves. In the case of valves, these are shown on the flow plans, and the appropriate pattern should be selected or the pattern adjusted for the actual number of valves. For the pipe, flanges, and fittings, the actual design will most likely vary from the estimate on a line-by-line basis but the overall totals should agree closely (i.e., the average quantities estimated for a particular design should be close to the actual design, assuming average performance by the designer).

Because any manual manipulation of the thousands of items mentioned earlier that make up a piping estimate would be a tedious operation, the company computerized its unitized piping method. Table 1 (page 44) is a typical

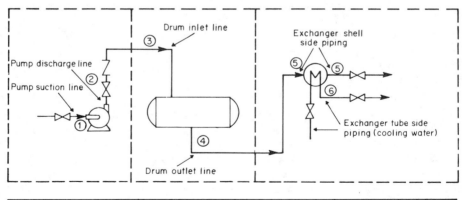

Line no.	Pattern type	Line no.	Pattern type	Line no.	Pattern type
1	01	3	70	5	30
2	02	4	71	6	43

Figure 3   How piping for simple process is broken down.

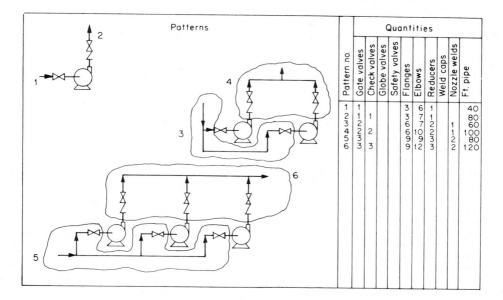

Figure 4   Typical pump–piping configuration.

printout of the line-by-line summary. The printout, in addition to including the input data and quantity results, also includes the costs of the materials (quantities times average unit costs), the fabrication cost (average number of welds per pattern times average vendor unit prices), and erection manhours (quantity times average manhour rates to erect).

Table 2 is another printout, this time the same data summarized in a different form. Instead of a line-by-line summary, the quantities have been segregated according to size, line rating, and materials of construction. This information is used to compare the estimate's prediction of the total quantities with the actual takeoffs as they are prepared by the contractor. In fact, in many instances these estimated quantities, which are arrived at much earlier than the takeoffs, are used as the basis for the initial purchase of piping bulk materials. This is especially true when it is known that pipe material delivery times are lengthy.

Once a method such as unitized piping is developed and put into use, it can be (and has been) further sophisticated and improved through feedback and additional development. For example, new patterns can easily be developed and added as the need requires. Pattern modifications can be made (adding a valve or flanges or length of pipe where it is known that average conditions will not

Table 1 Typical Piping Line-By-Line Printout

Location Netherla.						Facility est. by Program			Estimate ECEU7514				
					Line-by-line summary at location for rev out						Valves		
Line ident	Pat No.	Size inch	No ftg	Feet pipe	Mtl	Flng rate	Pipe schdl	GA	CK	GL	SV	OP	
S1 04	540	4.0	2	10	CS	150	40 C	0	0	0	0	0	
S1 94	541	4.0	2	4	CS	150	40 C	0	0	0	0	0	
S1 05	540	6.0	4	10	CS	150	40 C	0	0	0	0	0	
S1 95	541	4.0	2	4	CS	150	40 C	0	0	0	0	0	
S2 18	491	0.7	2	5	CS	150	80 I	0	0	0	1	0	
S3 12	540	4.0	2	10	CS	150	40 C	0	0	0	0	0	
S3 92	541	4.0	2	4	CS	150	40 C	0	0	0	0	0	
S4 12	504	1.5	2	7	CS	150	80 I	1	0	0	0	0	
S4 13	505	2.0	3	7	CS	150	80 I	0	0	0	0	0	
S4 14	396	1.0	4	10	CS	300	80 I	0	0	1	0	0	
00 03	191	10.0	22	1740	CS	150	20 C	1	0	0	0	1	
00 46	219	4.0	2	0	CS	150	40 C	1	0	0	0	1	
00 47	491	4.0	8	255	CS	150	40 C	1	0	0	1	1	
00 38	511	4.0	1	20	CS	300	80 C	1	0	0	0	0	
00 42	511	1.0	10	250	CS	300	40 C	1	1	0	0	0	
U0 12	511	12.0	0	150	CS	150	20 C	0	0	0	0	0	

Pipe Totals

| | Tons | | Material |
	Shop only	Total	DFL
Pipe ex steam trace	3.00	30.64	72027.
Additional small bore pipe		0.04	576.
Steam trace		0.16	858.
Accessories			2652.
Chemical cleaning			0.
Pipe total code 313.1	3.00	30.83	76113.
Misc. pipe support code 318		0.41	1718.
Insulation code 348			5600.
Paint code 349			16800.
Electrical code 337.1			9306.
Grand total	3.00	31.25	109538.

(continued)

Table 1 (continued)

| Revision | | | | Date 02/07/83 | | | | | | Process Pipes – Code 313.1 | | |
| Flngs | | | | Totals | | | | Steam tracing | | Insulation | | |
RG NO	OR PR	NOZ WLD	EL CNT	Tons	D FL	Fabrication	Erect hours	Mtl	Hrs	In	Matl D FL	Hrs
0	0	0	0.2	0.06	112	251 D FL	10	0	0	0.0	0	0
2	0	0	0.0	0.04	153	407 D FL	11	0	0	0.0	0	0
0	0	0	0.2	0.13	256	643 D FL	18	0	0	0.0	0	0
2	0	0	0.0	0.04	153	407 D FL	11	0	0	0.0	0	0
2	0	2	0.5	0.01	751	0	9	0	0	0.0	0	0
0	0	0	0.2	0.06	99	249 D FL	9	0	0	0.0	0	0
2	0	0	0.0	0.04	153	407 D FL	11	0	0	4.0	391	5
3	0	1	0.5	0.03	241	0	18	0	0	0.0	0	0
1	0	2	0.5	0.02	100	0	33	0	0	0.0	0	0
0	0	0	0.0	0.02	145	0	12	0	0	0.0	0	0
2	0	0	0.8	25.47	45321	4099 D FL	2014	0	0	0.0	0	0
2	0	0	1.0	0.09	5545	7 D FL	36	0	0	0.0	0	0
4	0	2	0.5	1.62	12897	473 D FL	222	0	0	0.0	0	0
2	0	1	1.0	0.28	1316	426 D FL	30	0	0	0.0	0	0
4	0	0	1.0	0.23	1077	0	108	525	106	2.5	5208	80
0	0	0	1.0	2.50	3708	575 D FL	223	0	0	0.0	0	0

For rev out

	Materials		Labor			
	Shop FAB D FL	Total D FL	Field FAB HRS	Erection hours	Total hours	Total D FL
	7945.	79973.	0.	2777.	2777.	111918.
		576.		33.	33.	1343.
		858.		130.	130.	5253.
		2652.				
		0.				
	7945.	84058.	0.	2941.	2941.	118514.
		1718.	0.	111.	111.	4956.
		5600.		85.	85.	0.
		16800.		329.	329.	0.
		·9306.		298.	298.	10667.
	7945.	117483.	0.	3764.	3764.	134137.

Table 2 Typical Piping Bulk Material Quantity Printout

Location Netherla.	Facility est by Program		Estimate ECEU7514	
	Bulk piping summary			
Material	Item	Size inch	Sch. or rating	Quantity	Unit
Carbon steel	pipe				
	pipe	0.5	40	271.	ft
	pipe	0.7	80	5.	ft
	pipe	1.0	40	250.	ft
	pipe	1.0	80	10.	ft
	pipe	1.5	40	25.	ft
	pipe	1.5	80	7.	ft
	pipe	2.0	80	7.	ft
	pipe	4.0	40	287.	ft
	pipe	4.0	80	20.	ft
	pipe	6.0	40	10.	ft
	pipe	10.0	20	1740.	ft
	pipe	12.0	20	150.	ft
	pipe	Subtotal			
Carbon steel	90 elbows				
	90 elbows	1.0	40	10.	each
	90 elbows	1.0	80	4.	each
	90 elbows	1.5	40	3.	each
	90 elbows	1.5	80	1.	each
	90 elbows	2.0	80	1.	each
	90 elbows	4.0	40	12.	each
	90 elbows	6.0	40	2.	each
	90 elbows	10.0	20	22.	each
	90 elbows	Subtotal			
Carbon steel	reducers				
	reducers	4.0	40	6.	each
	reducers	6.0	40	2.	each
	reducers	Subtotal			

Table 2   (continued)

Revision	Date 02/07/83	Piping — Codes 313.1 and 313.2

For carbon steel

Cost at location		Weight	
Unit price	Total	tons	Notes
			Including steam tracers
1.525	413.	0.11	Sum of all schedules in this size
2.933	15.	0.00	Sum of all schedules in this size
3.238	810.		
3.780	38.	0.22	Sum of all schedules in this size
4.720	118.		
5.769	40.	0.05	Sum of all schedules in this size
4.412	31.	0.02	Sum of all schedules in this size
8.149	2339.		
11.488	230.	1.70	Sum of all schedules in this size
14.376	144.	0.09	Sum of all schedules in this size
19.793	34440.	24.39	Sum of all schedules in this size
24.365	3655.	2.50	Sum of all schedules in this size
D FL	42272.	29.08	
			Including steam tracers
6.959	70.		Miter material cost with pipe
6.959	28.		
14.273	43.		
14.273	14.		
6.335	6.		
15.393	185.		
38.212	76.		
128.495	2827.		
D FL	3249.		
8.789	53.		
17.728	35.		
D FL	88.		

apply on your project). Where it is known that the plot plan layout is obviously not typical and not average, overall modifications can be made. In fact, in the case of the company that originated the method, it has now been developed to the point where the coordinates for each piece of equipment (location on the plot layout) can be used to develop a more accurate assessment of the average length of piping required to connect the equipment items to one another.

Since the advantages of unitized piping typify the advantages of estimating methods in general, it will serve a useful purpose to list these advantages:

Good (accurate) results
Fast
Consistent
Easy to learn
Easily modified
Ideal for computer
Excellent control tool
Strictly in-house (timing/manpower minimal)
Living system (fine-tuning)

METHODS FOR OTHER BULK MATERIALS

Similar methods can be developed for other bulk material items, such as concrete (already illustrated earlier), structural steel, electrical lighting and wiring, insulation and paint. For example, the quantity of steel for a typical pipe support can be developed from a method based on analysis of past design for pipe supports, separated into types (single-support, tee-support, bent or U-frame support, with adjustments for length between supports). In a similar way, a method can be developed for steel structures by relating tonnage required to the anticipated dimensions or volume of the frame of the structure, or to the load on the structure. Lighting quantities (number of fixtures) can be related to square footage of area to be covered by lighting. Insulation and paint can also be related to square footage or, in the case of piping, to number of feet of a certain size pipe.

All of these methods would be developed primarily from the company's past historical records or from published data by others, but adjusted to meet the needs and the specific manner in which the company executes a project (i.e., company standards, material requirements, performance of the specific contractors that the company does business with, etc.). Further, in all cases, an effective feedback system would be used to insure that the method is giving reasonable, realistic results and to adjust or fine-tune the method where this is not happening.

AVERAGING CONCEPT: SUMMARY

The averaging concept is a simple approach to methods development and provides a simple approach to estimating. It reflects past, known performance and as such gives consistent and reasonable results. It is easily maintained, computerized, and understood. As we will see under cost control, it provides a good yardstick; it also eliminates the need for preliminary engineering and vendor quotes, and can be applied to all types of estimates from curve through definitive estimates.

DEFINITIVE ESTIMATES: THE DETAILED APPROACH

Up until now, we have been discussing the averaging concept and the use of this concept in developing methods for preparing in-house definitive estimates (i.e., detailed definitive estimates that can be made wholly within the cost-engineering department of a company without assistance from the engineering or purchasing department and without reliance on vendor quotations). A second approach is an even more detailed one, where the estimate is based mainly on the specific requirements of the project rather than the average. Although there are estimating methods involved, for the CE these methods are more related to how to put the quantities together than the estimate of the quantities. In fact, for most items, the CE acts as a coordinator and the quantity estimating is done by others in the company or even outside the company, as in the case of vendor quotations.

Figure 5 illustrates the magnitude of the effort that goes into the preparation of a definitive estimate. Note the number of departments involved at the top of the figure (i.e., engineering, construction, procurement, etc.). It can readily be seen from the number of tasks and the number of people involved that preparing a definitive estimate is a genuine team effort and requires a considerable amount of coordination, most of which is done by the CE.

Although information in the form of quantities, vendor quotations, and execution plans are prepared to a large extent by others and fed into the cost-engineering group, the CEs have the responsibility of translating these data into the estimate of what it will cost to build the plant. Here judgement, experience, and training come to play as the CE must answer such questions as:

Is the single-vendor quote for a particular item that we have received competitive and representative of what we will eventually buy?

How much escalation must be added?

Based on past experience, what must be added to the preliminary quantity take-offs to render them complete and all inclusive?

How will the construction work be done: subcontract or direct hire?

What will be the labor productivity?

What contingencies must be added?

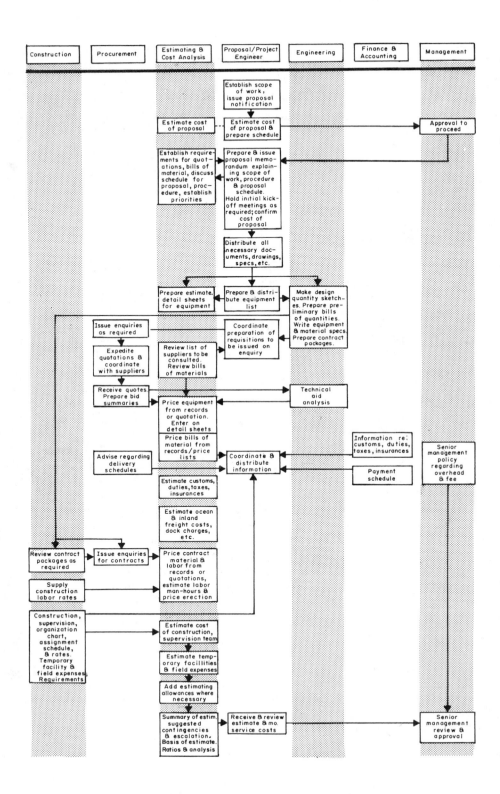

These and many other questions must be answered and adjustments made by the CE to the data received from others. In addition to coordinating, reviewing, and adjusting the data, the CE also has to do a considerable amount of estimating. All the field labor manhours are estimated by the CE based on the quantity takeoffs. The CE also estimates the engineering costs, FLOH (supervision and field expenses), and other items such as freight, duty, taxes, and insurance. In estimating these items, the CE will probably be using in-house methods developed by the contractor's estimating department along the lines we have been discussing earlier (i.e., an averaging concept approach).

Because preparing a definitive estimate represents such a large team effort by the contractor, an extensive review and checking process is involved before the estimate can be released and/or used as the basis for a bid. Because of the many steps taken and the many people and interfaces involved in the preparation of the estimate, this elaborate checking process is extensive but essential to insure that no misinterpretations, omissions, or gross errors have been made.

COMPARISON OF THE TWO APPROACHES

In this chapter we have been discussing two approaches to estimating methods and, hence, estimating. The first is the in-house method where all methods developed and used allow the estimate to be prepared wholly within the confines of the cost-engineering department, independent of other resources and departments. The methods developed are based on an averaging concept, using past historical data, preferably that of the company. This approach requires a significant initial effort and expenditure in developing the methods. A further effort is required to maintain and update the methods. This approach is generally used by clients or owners.

The second approach is the more detailed one used primarily by the contractors. This method tries to predict as accurately as possible the specific contractor's plan of execution for that particular project. Consequently, there is a minimum of the averaging concept and considerable reliance on preliminary engineering, layouts, takeoffs, and quotations on that specific project. Much of the effort and costs related to the preparation of the estimate are on a pay-as-you-go basis and a minor effort is expended in methods development.

Figure 5   Preparation of estimate flow chart.

ADVANTAGES/DISADVANTAGES OF EACH APPROACH

Let us examine briefly the advantages and disadvantages of the two approaches. The contractor's more detailed approach reflects his specific performance on that specific project and tends to be more accurate than the averaging approach. Also, since the contractor has prepared the estimate in a detailed manner, he has more confidence in it and will tend to be more receptive to its use as a control estimate on a reimbursible contract than any estimate provided by a client and based on an averaging approach. The detailed approach suits the average contractor's approach to financing and budgeting (i.e., have each project pay for itself rather than invest large sums of money in methods development initially, for future use).

On the other hand, the averaging concept or in-house approach enables the CE to prepare definitive estimates of satisfactory accuracy in far less time and with much less manpower. The estimates are available early in the project execution stage and, as we will see later, this is an important consideration in cost control. Further, by reflecting average performance instead of specific performance, the estimate is a good yardstick for the owner for measuring contractor performance and, hence, an excellent control tool.

It can be seen from the above advantages and disadvantages that the approach used depends to a large extent on management's decision to spend money in the area of methods development. A good case can be made for more in-house methods to be developed and used by contractors, despite the high initial cost. In-house methods can be made accurate, as accurate as detailed estimates, especially where the averaging concept is only for the specific contractor (vs. many contractors, as is the case when the owner develops his methods). Being able to complete estimates faster than the competition gives the contractor an edge; in addition, once the initial expenditure has been made, the cost of preparing future estimates drops dramatically.

This is significant, since the normal approach to preparing a contractor's definitive estimate can involve more than 20,000 manhours of total effort for a major project estimate. Further, with less manpower consumed on estimates, more is available for making new proposals. In-house estimates would also open the door to an area where few contractors have ventured [i.e., that of preparing early, conceptual-type estimates (screening and budget estimates)]. It would also make estimating more interesting and challenging to the contractor's cost-engineering staff, replacing pricing up of quantity takeoffs, and providing more time for making judgement-type decisions. In effect, under the right conditions, developing a complete in-house estimating methods approach could be a bonanza for a contractor.

ESTIMATING METHODS FOR INSTALLATION LABOR

All examples and discussions we have had relating to both the averaging concept (in-house) approach and the detailed definitive approach have concentrated on establishing the quantities of materials required. Obviously the material, once designed, procured, and delivered, has to be installed in the field. Therefore, methods also need to be developed for estimating the manhours required to do this installation.

The approach used to develop an estimating method for erection labor manhours is similar to the approaches already explained for material quantities. A heavy emphasis is placed on historical data and feedback from field experience on past jobs in the area where the project will be built. Unit rates are established for each category, for example, the manhours per ton for each category of steel (heavy structures, platforms, pipe supports, etc.), for paint and insulation, and manhours per square meter. The manner in which these data can be collected is covered in Chap. 7.

A key factor in developing a good estimate of erection manhours is labor productivity (i.e., for a given project built at a specific location and time, what is the best prediction of the final manhour rates that will be performed?). This subject has perplexed CEs, managers, and field supervisors and, because of the many factors involved, has a wide degree of variability. Because of its complexity and importance, this subject will be covered in depth separately in Chap. 5.

ESTIMATING SUBCONTRACT LABOR COSTS

In areas where experienced subcontractors are available, this approach to field erection is often followed instead of directly hiring labor from the local labor pool. Subcontracting is used almost exclusively in Europe and Asia and, to a lesser degree, in the U.S. and Canada. When it is known that the subcontracting route will be followed, the estimator requires a method for predicting the subcontract labor costs. Here, again, there are two possible approaches or methods—definitive and in-house. In the definitive approach, preliminary quotations are received from one or more subcontractors and the estimator will analyze the quotes and adjust, if necessary, to arrive at his prediction of the most probable final cost of the subcontract.

An in-house method would involve the development of average subcontract unit rates for each subcontract category (piping fabrication or installation, equipment erection, civil works, etc.) and then multiplying these rates times the quantities and/or manhours established by the other methods described earlier. The unit rates are data and will be discussed again briefly in Chap. 7 but are explained in detail here to complete the discussion on the estimating methods for installation labor.

There are several ways of developing average unit rates. Unit rates are often available from past labor only contracts or as separately quoted "day-works" (reimbursable labor costs) on lump sum or unit price subcontracts. These can be analyzed and average costs developed.

Another approach is the analysis of the final costs for a subcontract, using an estimate of the manhours consumed during the contract period. Table 3 is an example of such an analysis.

A third approach is to develop the required average unit prices by estimating the cost of each of the main components of the unit rate. This synthetic build-up approach is illustrated in Table 4. It should be noted that for an equipment or piping erection contract, the unit rate will be high, whereas for a low equipment intensive subcontract, such as instrumentation or electrical, the rate would be much lower.

The subcontract cost can vary from location to location and is affected by three main factors—the direct wage rate, the total field overheads, and the labor productivity. Many times these factors tend to be offsetting. Table 5 is an example which illustrates how major differences in wage rates can be almost entirely offset by differences in overheads and field productivity.

METHODS NEEDS vs. APPROACH USED

The degree of sophistication applied to the area of estimating methods by a company should be balanced by that company's needs. A large company will more likely need a full 100% capability of making estimates from the screening stage through the definitive phase. Further, the large company is most likely to able to afford the comprehensive effort required to set up and maintain the

Table 3   Subcontract Rate: Example

Quantity	Unit Price		Total Cost $
5 tons	Small bore @	$900/ton	4,500
20 tons	CS Piping (2" - 8")	$600/ton	12,000
10 tons	CS Piping (10" - 16")	$500/ton	5,000
5 tons	ST ST Piping (2" - 8")	$800/ton	4,000
	Height Extras/Stand-by Time		2,000
		Total contract	$27,000
Crew	20 men/duration 3 weeks		
	20 men × 40 hr/week × 3 week		
		2,400 hr	
Equivalent rate		27,500/2,400 = $11.46/hr	

Table 4   Subcontract Rate: Synthetic Buildup Example

	$/hr
Base hourly wage	5.12
Social/union benefits	1.30
Indirect labor (20%)	1.02
Equipment cost per manhour	1.75
Consumables/supplies	0.30
Supervision	0.90
Overheads/profits	1.20
	11.59

system required. Also a large company has more resources available to do the job and more historical data to draw from for developing its estimating methods.

Finally, many large companies have a strong preference not to make public the development of new projects. Having in-house methods and an in-house estimating capability avoids the necessity to go to outside vendors and subcontractors. On the other hand, a small company is working from a smaller margin, has less resources and less historical data. The small company does have a need to develop the capabilities of getting early estimates done (screening and budget estimates) but can and often does rely on contractors and suppliers for more detailed estimates once a project has been approved and goes into the project execution stage.

Table 5   Typical Construction Cost Variations

Location	Direct cost rate	Total overheads	Total field rate	Field productivity	Equivalent field rate
	$/hr.	%	$/hr.	%	$/hr.
A	2.00	425	10.50	50	21.00
B	10.00	125	22.50	90	25.00
C	20.00	20	24.00	100	24.00

A = Direct hire, underdeveloped country.
B = Direct hire, U.S.A.
C = Subcontract, Europe.

COMPUTERIZED ESTIMATING

There are a large number of benefits to be derived from using a computer to assist the cost engineer in the preparation of an estimate. Some of these benefits are the following:

Speed
Consistency
Fewer errors
More output
Neater presentation
Improved communications
Input form become a checklist

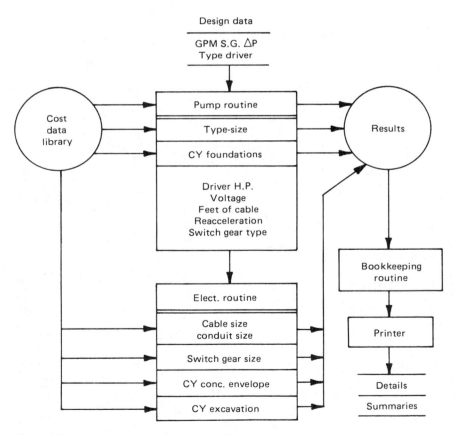

Figure 6   Schematic example of a detailed estimating program.

To use the computer, programs must be developed or purchased. Buying a program makes it available for use sooner, enables the user to use the experience of others and, overall, costs less. On the other hand, the program purchased may not meet your needs and it may be difficult to verify the quality of the program. Further, it may be difficult to revise. All of these factors must be considered when deciding whether to buy or to develop programs.

Computerization can be especially useful to the estimator in the preparation of semidetailed estimates. There are a number of commercial programs available which use equipment physical size or a similar parameter and, from this information, generate bulk material quantities and costs, based on historical averages. This was discussed briefly in Chap. 3. These programs can often be adapted to suit a company's in-house needs.

Figures 6 and 7 are schematic examples of a detailed and a semidetailed estimating program. Table 6 is an example of a typical printout for electrical power cable and pushbutton wiring. Earlier in the chapter, Tables 1 and 2 were given as examples of typical piping printouts.

SOME GUIDELINES FOR METHODS DEVELOPMENT

In developing an estimating method, experience has shown that the following guidelines should be followed:

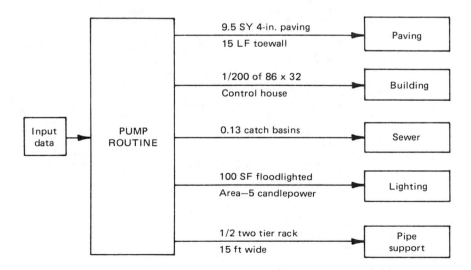

Figure 7 Schematic example of a semidetailed estimating program.

Table 6 Example of a Typical Printout for Electrical Power Cable and Pushbutton Wiring[a]

| | Location Baytown Facility | | Estimate | Revision | | Date 02/12/74 | | Electrical – Code 337.1 | | |
| | | | | Power Cable | | Power Cable Conduit | | Pushbutton | | Wiring 1" |
Identification	Motor HP	Volt KV	Run ft	Size	Matl cost	Size in.	Matl cost	Size	Matl cost	conduit cost
BR CRANE	5.0	0.600	200	12	37	1.0	130	14	52	0
BR CRANE	5.0	0.600	200	12	37	1.0	130	14	52	0
BR CRANE	5.0	0.600	200	12	37	1.0	130	14	52	0
C-301A/B	250.0	9.000	200	2	386	2.0	270	14	83	130
C-301A/B	250.0	9.000	200	2	386	2.0	270	14	83	130
E-306	2.0	0.600	200	12	37	1.0	130	14	52	0
E-306	2.0	0.600	200	12	37	1.0	130	14	52	0
E-308	15.0	0.600	200	10	55	1.0	130	14	52	0
E-308	15.0	0.600	200	10	55	1.0	130	14	52	0
E-312	2.0	0.600	200	12	37	1.0	130	14	52	0
E-312	2.0	0.600	200	12	37	1.0	130	14	52	0
P-301A-C	3.0	0.600	200	12	37	1.0	130	14	52	0
P-301A-C	3.0	0.600	200	12	37	1.0	130	14	52	0
P-301A-C	3.0	0.600	200	12	37	1.0	130	14	52	0
P-302	0.5	0.600	200	12	24	1.0	130		0	0
P-303A/B	15.0	0.600	200	10	55	1.0	130	14	52	0
P-304	0.5	0.600	200	12	24	1.0	130		0	0
P-305A/B	100.0	0.600	200	2/0	395	2.5	427	14	52	130
P-306A/B	250.0	9.000	200	2	386	2.0	270	14	83	130
P-307	2.0	0.600	200	12	37	1.0	130	14	52	0
P-304	3.0	0.600	200	12	37	1.0	130	14	52	0
MIX-305	1.0	0.600	200	12	37	1.0	130	14	52	0
MIX-306	1.0	0.600	200	12	37	1.0	130	14	52	0

| Identification | Switchgear | | | | | MTR S/G Term cost | SPL KITS A/G PULL points cost | Total matl USDL | Total labor hours | Total labor USDL |
	Break or fuse	RE-ACC	IND OR OUT	MVA RTG	Matl cost					
BR CRANE	BRK	NO	IND	0	345	54	0	618	67	635
BR CRANE	BRK	NO	IND	0	345	54	0	618	67	635
BR CRANE	BRK	NO	IND	0	345	54	0	618	67	635
C-301A/B	BRK	YES	IND	500	18742	88	0	19747	181	1705
C-301A/B	BRK	YES	IND	500	18742	88	0	19747	181	1705
E-306	BRK	YES	IND	0	465	54	0	738	67	635
E-306	BRK	YES	IND	0	465	54	0	738	67	635
E-308	BRK	YES	IND	0	533	54	0	823	75	711
E-308	BRK	YES	IND	0	533	54	0	823	75	711
E-312	BRK	YES	IND	0	465	54	0	738	67	635
E-312	BRK	YES	IND	0	465	54	0	738	67	635
P-301A-C	BRK	YES	IND	0	465	54	0	738	67	635
P-301A-C	BRK	YES	IND	0	465	54	0	738	67	635
P-301A-C	BRK	YES	IND	0	465	54	0	738	67	635
P-302	BRK	YES	IND	0	248	24	0	425	59	557
P-303A/B	BRK	YES	IND	0	533	54	0	823	75	711
P-304	BRK	YES	IND	0	248	24	0	425	59	557
P-305A/B	BRK	YES	IND	0	1230	122	0	2155	157	1482
P-306A/B	BRK	YES	IND	500	18792	88	0	19747	181	1705
P-307	BRK	YES	IND	0	465	54	0	738	67	635
P-304	BRK	YES	IND	0	465	54	0	738	67	635
MIX-305	BRK	NO	IND	0	345	54	0	618	67	635
MIX-306	BRK	NO	IND	0	345	54	0	618	67	635

[a]Motor and switchgear termination material costs include fittings such as 111 and 1 grommets, sealing bushings, connectors, tape, stress cone kits when required, pushbutton station supports, and short lengths of conduit to complete the connections as necessary. Quantities and costs do not include waste allowance.

61

Keep the approach and method simple. The more complex, the less likely it will be used and the greater the opportunity for error.

If the method must be complex (like piping), use the computer.

Bear in mind that the primary function of an estimating method is to define hardware (quantities).

Keep methods and data separate. They are two different functions.

Make the method all-inclusive and avoid ending up with vague estimating allowances to make up for omissions.

Test the method results for accuracy and reasonableness.

Set up and maintain a feedback system to fine-tune the methods after development.

If at all possible, assign full-time CEs to methods development.

5

LABOR PRODUCTIVITY

GENERAL

In Chap. 4, we discussed briefly the development of an estimating method for erection labor manhours. We also emphasized that a key factor in developing a good estimate or erection workhours is field labor productivity. In this chapter, we will investigate why labor is so important in an estimate, the establishment of a base productivity for an area, the many variables that can affect this base productivity, and the approach to be used in developing an estimate of the labor productivity for a given project.

THE IMPORTANCE OF LABOR IN AN ESTIMATE

Estimating or predicting the cost of labor for a given project is, like estimating piping costs, a difficult task that has frustrated both CEs and field supervisors. The cost of labor is determined by multiplying manhours (sometimes called workhours) by the applicable wage rates (i.e., multiplying quantities by unit costs). How one establishes the correct wage rates to be used is discussed in detail in Chap. 7. Establishing the manhours can only be accomplished after establishing or predicting the manhour rate (or productivity) that will be achieved on the project. Direct labor constitutes a high percentage of the total cost of a project (an average of 15%, with a fairly wide range, depending on the type of project and the location). Since this cost and the associated schedule is affected directly by the labor productivity achieved, it is of vital importance to CEs and to the accuracy of their estimates that the correct labor productivity value be used in the estimate. Unfortunately, field labor productivity is the single greatest variable in any estimate and is extremely difficult to estimate. Often, estimators will not try to do more than use the simplest approach since they feel there is no real science to predicting labor productivity. As a result, many estimates of labor manhours are overrun in the field, often with a disastrous impact on the project. Not only does the final cost of the project exceed the appropriated amount but, because of the close relationship that exists between erection manhours and the project schedule, more than likely the completion date of the project is extended or delayed, with the associated debits of not meeting the project objectives in the area of production and marketing.

THE SCIENTIFIC APPROACH

Because labor productivity is a function of a number of specific variables, and because these variables can be assessed beforehand and their impact on productivity predicted for a specific project with some degree of accuracy, estimating methods and approaches can be developed along more scientific lines than just using the manhour rates from the last completed project. This type of "scientific approach" not only produces a more logical accurate estimate of the manhours required for a project but, because it is an estimating method, it carries with it all the inherent advantages of estimating methods in general, as outlined and discussed in Chap. 4. A more comprehensive approach coupled with the better understanding of the labor productivity variables also contributes significantly to the analysis of the actual performance in the field and, hence, to field cost control, as will be discussed in Chap. 21.

ESTABLISHING BASE PRODUCTIVITY

The first step in establishing a method for estimating labor productivity is to establish the base or native productivity for the area in which the project will be built. Base productivity represent the manhour rates that can be expected at a location under "normal" or "average" conditions.

The criteria to be used in establishing this normal or average condition will become obvious as we explain how to arrive at the base productivity. First, let us examine those items or variables that affect base or native labor productivity:

Local climate
National characteristics
Local craft availability
Labor mix (local vs. expatriate)
"Union atmosphere"
Degree of equipment use
Degree of supervision

All of these items are factors that have contributed to the establishment of a recognized labor productivity for a given country, and, in the case of large countries, for a given sector of the country. They are factors or variables that under normal circumstances tend to be fixed for fairly long periods of time. For example, it is generally accepted that manhour rates in tropical climates will be lower than those in the more temperate zones. In less developed countries, there is often a minimum of equipment use and manhour rates to carry out such functions as excavation can be high. Many countries, such as Japan, have as a national characteristic an industrial approach with high labor productivity as a base. Consequently, over the years, because of these factors, a base productivity level can be established for a given location.

How then does one establish this base productivity? Some information can be derived from published data, especially from local government labor agencies. Unfortunately, this data source is meager and difficult to analyze and apply to a specific company's needs in a given industry. A better approach is to use historical data on past construction projects. These data have to be adjusted for "abnormal" conditions, as we will explain later in this chapter. The base productivity arrived at in this manner can be kept current through labor surveys conducted by the company's CEs and construction people at regular intervals. If the company has had no prior experience in a given location, then a special labor survey would be conducted to secure the necessary information.

During these surveys, local contractors, subcontractors, government agencies, and local affiliates of the company would be contacted for the latest picture on labor productivity. Any long-term deterioration or improvement can then be fed back into the estimating method and the base labor productivity figure adjusted accordingly. This survey is also used to gather other data, such as short-term abnormal conditions affecting labor and labor escalation. These are discussed in more detail later in this chapter and in Chap. 7.

WHY ADJUST BASE PRODUCTIVITY?

If normal or average conditions existed all the time, we could always use the base productivity for each of our projects. Unfortunately, there are almost always at least one or two abnormal conditions related to a specific project and these must be considered if we are to arrive at a reasonable prediction of the manhours required for the project. Base productivity may have to be adjusted for one or more of the following:

Specific work conditions
Amount of overtime
Job size
Learning curve effect
Labor availability ("activity")
Contracting approach (subcontracts vs. direct hire)

Each of these can have a significant impact on labor productivity and, when combined, can have a multiplying effect. Let us look at each individually.

Specific Work Conditions

The physical conditions of a specific plant location can effect the labor productivity if these conditions are abnormal. For example, if the project involves modification to an existing unit (normally referred to as a "revamp" or "debottlenecking" project), the conditions under which the work will be performed are far from ideal and certainly not normal or average as the case would be with a new plant. Consequently, labor rates and labor productivity have to be adjusted accordingly. Productivity decreases because of such items as nonoptimum erection sequence, interferences from existing equipment, restrictions on "hot work" for safety reasons, and site accessibility. Each of these can be analyzed, based on past experience, and adjustments made. The area of revamp work alone can be responsible for a decrease in base productivity of as much as 30 to 40% and because of its importance, is discussed in more detail in Chap. 10.

Union negotiations occurring during the course of construction can and do have have an affect on productivity. Where it is known that contracts will expire during the construction period some adjustment should be made to productivity to reflect a predictable degree of labor unrest. Where it is known that local craft practices (jurisdictional requirements, etc.) and government regulations (safety, training, etc.) will most likely change during the course of the project schedule, appropriate adjustments to productivity for that project should be made. Where these changes in practices or regulations are of a permanent nature, then a change also needs to be made to the base productivity.

Impact of Overtime

It is common knowledge that efficiency or productivity drops off as the number of hours worked increases. Part of the basis for the estimate is a project execution plan and a predicted or planned work schedule and working hours. If these working hours are the normal pattern for the local area, then no adjustment need be made to the base productivity. If, however, it is known that overtime will have to be worked to meet schedule or to attract labor to the area, then adjustments must be made. Figure 1 illustrates the loss of productivity as both the number of hours and the number of days per week increases. This same loss is shown in Fig. 2 but here the loss is expressed as a factor adjustment to productivity as the percentage of manhours worked increase. The Business Roundtable recently completed a task force report on productivity which showed similar results as indicated in Fig. 3. With a known work schedule, CEs can reflect in their estimate the additional manhours required to work that schedule if

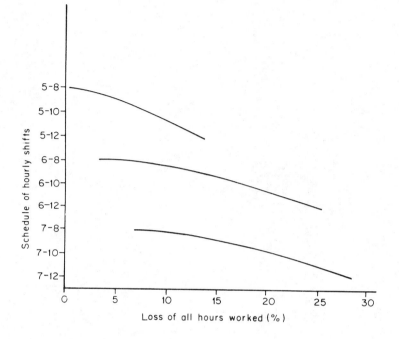

Figure 1 Overtime vs. productivity. (Date compiled from NECA summary, MCAA studies, statistics from U.S. Department of Labor Bulletin, no. 917, and other similar data.)

overtime is involved. This relationship of overtime vs. productivity is an important consideration also in cost and schedule control in the field when a construction manager is faced with the option of working overtime to gain schedule time. This alternative can be a costly venture with little gain and needs to be analyzed properly before a final decision is made.

Job Size vs. Productivity

Job size, in terms of the estimated number of manhours required to complete the project, can have a significant impact on labor productivity, as illustrated in Fig. 4. This figure assumed that the job size for which the base productivity was analyzed was 300,000 manhours (consequently, the productivity at this size is 100% or 1.0). The figure shows how, as the job size increases above 300,000, the productivity decreases and conversely, as the size decreases, the productivity improved. As projects become larger, they become more complex. Top key supervisors are spread thin, and lines of communications become longer, and the

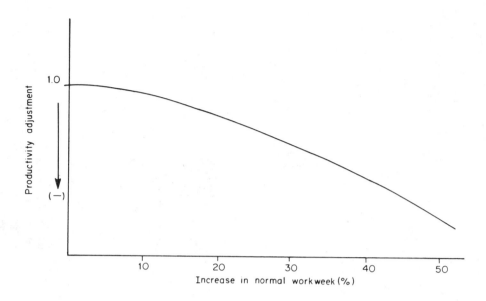

Figure 2   Overtime vs. productivity.

logistics of keeping workers and material at the proper level at the proper times become more difficult. The net result is a less efficient and less productive work force. The reverse is true as the job becomes smaller, simpler, and more easily managed.

Learning Curve vs. Productivity

Figure 5 illustrates the improvement in productivity that can be achieved through the "learning curve" effect. The dotted curve extension is the more typical, normal, bell-shaped manhour distribution curve for a project and is what one would expect to be the basis for the base productivity. However, an abnormal condition that can occur is to have a project in which initial planning and schedule analysis will show a manpower limitation (i.e., the normal peak manpower requirements cannot be met because of some constraint, such as available craft manpower being limited or known site conditions such as accessibility limiting the maximum number of workers). In these cases, often the only solution is to extend the schedule as shown in Fig. 5. However, offsetting this schedule debit is the credit to productivity that results from the extended peak period and its associated learning period. This situation has recently become

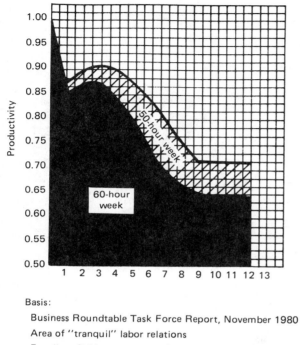

Basis:
Business Roundtable Task Force Report, November 1980
Area of "tranquil" labor relations
Excellent field management direction
Productivity = actual hours vs. standard base; measure weekly

Figure 3   Business Roundtable task force report on productivity.

more common as projects have increased in size and scope and have more frequently become manpower-limited.

Activity vs. Productivity

One of the most significant factors affecting base productivity is the degree of economic "activity" existing in an area during the construction period. If activity is high and there are many projects built in the area at the same time, the demand for labor will increase dramatically. This demand will have a twofold impact on local base productivity. First, all of the available local labor will be fully employed. In "soaking up" all the local labor, the project has acquired a number of marginally qualified craftsmen and helpers. These, when mixed with the average labor force, will dilute or reduce overall productivity from the levels attained during times of normal economic activity, the basis for our base productivity. Second, because there are so many job openings available, "turnover"

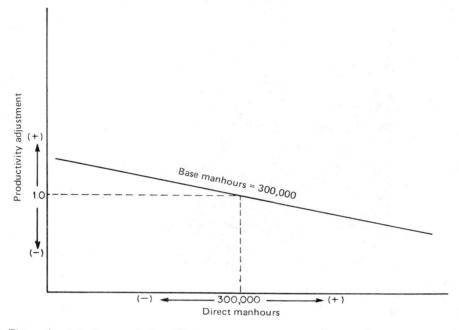

Figure 4   Job size vs. productivity.

(i.e., craftsmen quitting one job site to go to another one in the immediate local area) increases dramatically and this also creates a disruptive "auction" atmosphere, and this results in inefficiences and lower productivity.

Normally, during times of high economic activity, even the entire local labor force is insufficient to satisfy the demand and labor must be brought in from outside the local area. Experience has shown that when this happens, overall productivity will most likely decline, including the productivity of the local craftsmen. The reasons for this are not clear, since often the imported craftsmen are of average or better quality. The drop in productivity can most likely be attributed to one of attitude (i.e., the importing indicates that demand has exceeded supply, a situation that is almost always accompanied by lower productivity).

Figure 6 is a graphic illustration of the dramatic impact an increase in economic activity can have on the productivity of a project. The figure relates the ratio of the number of workers required in an area to the normal number of workers to a factor that can be used to adjust the base or normal productivity.

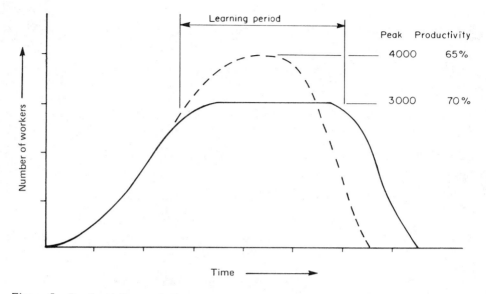

Figure 5   Productivity vs. the learning curve.

Direct-Hire vs. Subcontract Labor Productivity

There are essentially two ways a prime contractor can erect a plant in the field: either by hiring all the craftsmen and supervising them (the direct-hire labor approach) or by giving each identifiable piece of work to a specialist subcontractor, such as a pipe erection subcontractor or an instrument installation subcontractor (the subcontract labor approach) or, of course, a combination of both.

For a number of reasons, on the average, the subcontract labor will perform at an average 5 to 10% higher productivity level than the direct-hire labor. Most subcontractors have a close-knit and highly experienced work force, the nucleus of which probably has been working for the same subcontractor or at least in the same craft area for many years. Consequently, they work more efficiently. The same is true of the subcontractor's supervision. Also, subcontractors are experts in their fields or endeavor, say, pipe erection, whereas prime contractors are overall construction experts and do not tend to have the same degree of expertise in each specific area that specialist subcontractors have.

Although the end result is a high productivity for the subcontract labor route, it does not necessarily mean a lower labor cost or an overall lower project cost. The local area must have sufficient competent subcontractors to allow good competitive bids to be received (most subcontracts are a form of fixed-price contracts). Also, subcontracting creates some duplication in supervision,

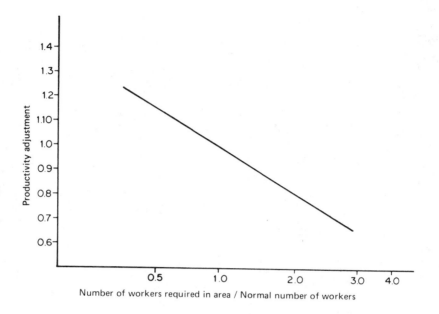

Figure 6   Productivity vs. "activity."

temporary facilities, and construction equipment and tool costs and these have to be included in the estimate. The decision to go subcontract or direct-hire can only be made after factors such as these are considered and analyzed.

ENGINEERING PRODUCTIVITY

The productivity of the designers, draftsmen, and engineers in the contractor's engineering office is also subject to variation depending on the geographic location, the contractor, and the project conditions. There is probably not the potential range of variability that we have been discussing for field labor productivity but it is a factor that can be significant enough for the estimator to consider. Variations of 10 to 15% are not uncommon.

Base productivity for a given area or a given contractor can be established in the same manner as was done with field labor productivity (i.e., by using historical data on past projects and adjusting for any abnormal conditions). It should be recognized, however, that the productivity of individual contractors cannot be compared directly without first analyzing the extent to which the contractor normally "engineers." For example, pipework can be engineered only down to, say, 3 in. in diameter, with smaller-size piping being "field run," without the benefit of detailed engineering. Another contractor may feel it is more economical overall to engineer *all* piping in detail to allow complete prefabrication and minimize field runs. In the latter case, the engineering manhours used will be more but the productivity will not necessarily be less. This must be considered by estimators when they make their productivity analysis.

Some of the abnormal factors that can affect engineering productivity are similar to those affecting field labor productivity. Overtime, job size, and activity all can have a significant impact on engineering productivity for the same reasons they affected field labor productivity. Specific working conditions (i.e., such as office layout and number of office buildings used) can also affect the overall productivity. All of these factors will be considered by estimators when they prepare their estimate or prediction of the engineering manhours required for the project.

SUMMARY

Some of the more important points discussed in this chapter follow:

Labor constitutes 15% of total project cost, has a large impact on schedule, and justifies a major effort in estimating.

Estimating or predicting labor productivity for a specific project is a difficult but possible task, providing a reasonably scientific approach is used.

The first step is to establish the base productivity for an area.

Base productivity is established by known conditions, such as local climate, national characteristics, and local craft availability.

Base productivity level is best established by conducting local surveys and from previous project feedback data.

Base productivity must be adjusted for known or predicted abnormal conditions for the specific project estimated.

Aside from specific physical work conditions, the amount of overtime, the job size, degree of activity, and the contracting approach are factors that can create abnormal conditions requiring adjustment.

Having considered all of the above items, estimators arrive at the productivity level for the specific project by multiplying the base productivity for the area by each of the adjustment factors that they have calculated for the abnormal conditions discussed in this chapter. This productivity is then applied to the unit erection rates, discussed in Chap. 4, to arrive at the labor manhours required for the project.

6

INDIRECT COSTS

GENERAL

Most of our discussions have centered on estimating or predicting the quantities and costs associated with the material and labor involved in a project; in estimating terminology, the *direct* costs. These are the costs of material items that are directly incorporated in the plant, such as a pump, drum, tower, or heat exchanger. Also included as direct materials are the bulk material items needed to complete the installation of the equipment item, such as its foundation, piping, and instrumentation. In this chapter we will cover the estimating of *indirect* costs (i.e., the quantities and costs of items that do not become a part of, but are a necessary cost involved in, the design and construction of the plant).

Indirect-cost items consist of a wide variety of items, but generally they fall into one of the following categories:

Detailed engineering
Contractors' fees
Temporary construction facilities
Construction consumables
Field supervision
Construction tools
Labor payroll burden
Miscellaneous costs such as insurance, freight, duties, and taxes
Owner costs

The technique used to estimate and control these costs is different than that used for direct costs, but the basic principles remain the same; we are still trying to predict as accurately as possible the manner in which the project will be executed. In the case of indirect costs, this is even more of a challenge to the estimator than direct costs, since many of the indirect-cost items are subject to a higher degree of variation than most direct costs. Let us examine briefly the approaches and techniques we can use to estimate some of these indirect-cost items.

DETAILED ENGINEERING

Detailed-engineering costs associated with a project normally amount to about 10% of the total project costs, with a broad range, depending on the type of plant and the size of the plant. In addition, the contractor's fee or administrative overhead and profit cost usually averages about 5% of total cost and, to a large degree, is determined by the amount of detailed-engineering and field supervisory services provided on the project. Consequently, detailed engineering is a significant cost item on a project and warrants the effort it requires to develop proper estimating methods and to prepare accurate estimates. Not only are the costs significant, but the engineering manhours and the engineering schedule that CEs must develop to prepare their estimate become a critical factor in controlling the overall schedule for the project.

Before discussing the scope of these detailed-engineering costs, we should mention that, especially for earlier estimates, it is necessary to predict the amount and cost of the basic engineering that must be done prior to the start of detailed engineering. Most of the time, this basic-engineering cost is an owner cost item and can vary considerably because of the many different and wide-ranging approaches used by different owners. Consequently, it is not possible to easily develop a method for predicting these costs. The following factors must be considered by CEs when estimating basic-engineering costs:

Type of process involved
Number of specifications involved
Complexity of each specification (i.e., degree of detail that will have to be
  provided to the contractor and, hence, to be included in the specification)

Often it is better to have the process department head make the estimate of the technical manhours required with CEs adjusting this estimate (usually upward) to reflect actual historical experience.

As discussed in Chap. 1, detailed engineering is based on the results of the basic-engineering or design work and can commence immediately after completion of the basic design. Detailed engineering, which follows and is done in a prime contractor's office, consists of the following scope:

Design and drafting
Project engineering and engineering specialists
Project management

Project control (scheduling and cost control)
Procurement
Expediting and inspection
Model work
OTT (other-than-time costs: telephone, telexes, reproduction, and computer)
Departmental overhead costs

With the exception of the last three items, all of the above-mentioned categories involve technical manhours only. Often, these manhours and their associated costs are referred to as home-office costs to distinguish them from field costs. Table 1 shows a typical breakdown of the cost of these manhours for a project. It is readily apparent from the breakdown that design and drafting costs are the most significant item of technical manhour cost and are a key factor in determining total detailed-engineering cost. Consequently, most of the methods developed for predicting engineering costs are based on first determining the amount of design and drafting manhours required and then relating the remaining costs to this item. Design and drafting manhours and costs are referred to as "production engineering," since a product (i.e., a drawing, specification, requisition) is a result of this effort.

SOME SIMPLE ESTIMATING METHODS

Figure 1 illustrates the simplest of methods that can be used to determine total engineering cost. This approach can be made more sophisticated and more accurate if the necessary basis information is available by introducing a family of curves covering different process plant types or one for on-sites and one for

Table 1   Typical Engineering Cost Breakdown

Area	%,$
Proposals	1
Design and drafting	58
Project management } Project engineering }	19
Procurement	12
Process engineering	3
Project control	3
Estimating	2
Construction planning	2
	100

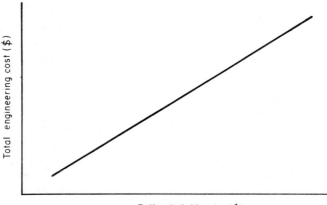

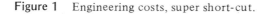

Figure 1   Engineering costs, super short-cut.

off-site engineering costs (i.e., one for the main process units, on-sites, and one for the support or off-site facilities).

Figure 2 plots the relationship of material cost and design and drafting costs, expressed as a percentage of material cost. A similar curve can be plotted of drafting manhours instead of costs. Table 2 is an example of a complete engineering cost estimate, starting with a simple curve. There are a number of points in the example worthy of mention:

All other technical manhours were determined by a factor relationship with drafting manhours, a common practice because of the consistent relationship.

To increase the accuracy of the estimate, the estimator used different hourly rates for the draftsman than the other, higher salaried engineers.

Departmental overheads were added at 115% of technical manhour costs. This figure can range from 75 to 150% and cover such costs as rent, depreciation, furniture, nontechnical and nonreimbursable manpower, such as secretaries and accounting personnel.

OTT costs were added at so much per manhour ($1.75), a common approach to predicting OTT costs.

THE UNITIZED APPROACH

Another simple estimating method used for determining the cost of engineering is a unitized approach. Here, based on historical data, drafting manhours are assigned to each of the major equipment items. The total drafting manhours are then arrived at by multiplying each equipment item by its unit manhours and

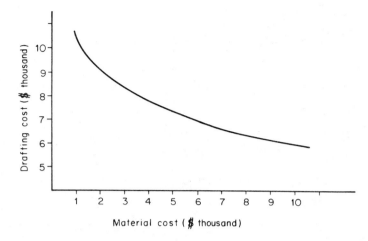

Figure 2   Drafting cost vs. material cost.

adding all the units together. To arrive at a more accurate total, the unit rates or the total can be adjusted to reflect the job size and complexity (deviations from the averages upon which the unit rates were based). Having determined the total design and drafting manhours, the CE then can estimate the total engineering cost by adding the other costs, as we discussed in Table 2.

A MORE DETAILED APPROACH

Although the simple methods described above are suitable for early preliminary and budget estimates, more detailed estimates are required for the definitive and control estimates required during project execution. These can take the form of a semidetailed approach or a completely definitive estimate.

A semidetailed approach that is used involves first establishing the piping drafting manhours required and then relating all other costs to this. The piping drafting manhours is established by counting the number of piping lines on a flow plan (or using an average number of lines per piece of equipment) and then multiplying these lines by a unit manhour rate (i.e., 25 manhours per line). Another approach is to establish from the number of piping lines the number of piping drawings to be made and multiply this by a unit rate of, say 300 or 400 manhours. In either case, however, adjustments must be made for abnormal conditions such as high temperature or high pressure, which can increase the number of design and drafting manhours because of potential layout and stress problems. From the total piping drafting manhours, the CE can estimate from past historical data the total design and drafting manhours required (piping is

Table 2   Quickie Engineering Estimate

Drafting manhours (from curve)	=	10,000
Other manhours (75%)	=	7,500
$ = 10,000 manhours × $10.00 ±	=	$100,000
7,500 manhours × $15.00 ±	=	$112,500
		$212,500
Overheads ± 115%	=	244,500
Other than time costs (OTT):		
17,500 manhours × $2.50	=	44,000
Total	=	501,000
Call	=	$500,000

usually about 33 to 50% of total drafting manhours). From there, the estimator can arrive at total cost as described earlier (Table 2).

The definitive approach involves a detailed look at each of the engineering categories. Drafting manhours is usually estimated by the head of the drafting department after a complete review of the work involved. Where possible, particularly in the area of project management, project engineering, engineering specialists, and project control, bar chart schedules are prepared showing the specific number of man-months for the individuals predicted to be assigned to the project. A separate estimate or quotation is received for the cost of preparing a scale model. OTT costs are estimated individually by item rather than using an overall cost per manhour. Overheads, however, are always added as either a percentage of hour costs or as a unit rate per manhour, since they are quoted to the client in this manner.

FIELD INDIRECT COSTS

A second significant area of indirect costs is commonly referred to as "field labor overhead" costs. These costs, like engineering, are also significant, amounting to 10 to 15% of total project costs. Field labor overhead (FLOH) consists of the following:

Temporary construction and consumables
Construction equipment and tools
Field supervision
Payroll burden

These items are identified and estimated separately for a project using direct-hire labor. In the case of subcontracts the indirect items would be included in the subcontract rate. In estimating any indirect item, there is certain critical basic information that significantly affects the cost of the item and, at minimum, this information must be available or assumed before a reasonable estimate can be prepared. The information required is the following:

Job size (usually in terms of the direct-labor manhours required)
Length of the construction period
The type of job—a plant or a modified existing plant
Local labor and construction practices

For most estimates, this type of information is available at the time of the estimate or can be developed with little effort. Let us examine briefly the scope of each of these categories of FLOH and some of the estimating approaches that can be used in arriving at their cost impact on a project.

TEMPORARY CONSTRUCTION AND CONSUMABLES

This category involves the material and labor associated with all the temporary work that must be installed to support the construction group and all the consumable materials involved in errecting the plant. Some examples of the items included in this area are:

Temporary buildings	Scaffolding costs
Temporary utility lines	Clean-up costs
Material handling and storage	Fuels
Temporary roads, fences	Welding rod
	Protective clothing

There are four basic approaches to estimating temporary construction and consumable costs. A "super short-cut" approach is to use a percentage of direct-labor cost based on past history. Another "quickie" approach is to develop a curve relating temporary construction costs to labor costs. A semidetailed approach would involve individual curves for each item, relating either the item quantity or size or the item costs to, again, labor cost. Finally, the detailed approach is based on making a layout and tabulation of the requirements from an execution plan and preparing an estimate of the material quantities, manhours, and costs involved. An important factor to remember, however, is that adjustments must be made when the project is not completely new. On plant additions, revamps, and debottlenecking projects, credit must be allowed for the fact that some of the permanent facilities and utilities can be used in lieu of the temporary facilities.

CONSTRUCTION EQUIPMENT AND TOOLS

Construction equipment and tools includes the following items:

Major construction equipment (cranes, bulldozers, etc.)
Minor construction equipment (welding machines, air compressors, etc.)
Hand tools (hammers, saws, wrenches, etc.)
Purchased and rented items (rental costs)
Rigging costs
Maintenance and repair costs
Freight costs

Job size and job length in the field are the two key factors that determine the cost impact of this area. Other factors are local equipment rates (how remote is the field site?), local practices (degree of emphasis on use of equipment vs. a manual approach), the number of heavy lifts and, finally, whether the equipment items will be purchased or rented. There are at least three approaches that can be used to estimate construction equipment costs. As was the case with temporary construction cost, a "super short-cut" approach can be used relating the cost to a percentage of direct-labor cost based on past history. A slightly more sophisticated way is to use a quickie curve, relating construction length or time to the total equipment costs.

The detailed approach required for the definitive and control estimates involves a prediction of the equipment needs for the project. Here a detailed list of each equipment item and the length of time it will be required in the field is prepared. These are then multiplied by the appropriate rental rate to arrive at the total cost. In all cases, it is extremely important that the CE adjust for any abnormal schedule conditions and for the need for any abnormally heavy lifts. The later item in particular can create the need for the rental of specialized and therefore high-cost lifting equipment, with a corresponding large impact on the estimate.

FIELD SUPERVISION

Field supervision includes both the contractor's permanent staff supervision and all locally hired supervisors and administrative and clerical staff. It specifically excludes craft labor and their immediate foremen since these are part of direct-labor costs. Included in the cost of field supervision are the wages of all the field office personnel, their payroll burden, any local living and travel allowances, and relocation costs. Also included would be the cost of operating the field office (i.e., telephone, duplication costs, furniture rental, etc.). The key factors that determine field supervision costs are the number of field labor manhours and the overall field schedule length. Other important factors are the site location (i.e., quality of the local field labor), the need for expatriate or foreign supervision, and the work schedule, particularly if shift work or extensive overtime is anticipated. Here again a "super short-cut" method can be used by relating field supervision costs to a percentage of direct-labor cost. A quickie curve can be used to relate construction length to total field supervision costs. For a definitive estimate, a detailed field supervision force list can be predicted along with appropriate salary rates and length of time on the job.

PAYROLL BURDEN

Direct-field labor payroll burden costs are more difficult to predict than any of the other FLOH items and consequently more difficult to estimate. Normally, the scope of payroll burden is considered to consist of the following:

Statutory benefits (social security, worker's compensation, insurance plans, etc.)
Public holidays
Paid vacations
Lost time due to illness
Training costs
Lost time due to weather
Other nonproductive time

The key factor needed for estimating the cost of payroll burden is an understanding of local statutory regulations and local labor agreements. Other variables that must be predicted are the amount of nonproductive time and need for training. Nonproductive time resulting from such factors as weather and sickness can be estimated based on historical data. Nonproductive time can increase sharply on a project involving work within an existing and operating unit because of such items as work stoppages due to unsafe conditions and smoking breaks. These must be considered and included in the estimate.

The simplest approach to estimating payroll burden is to relate it to a percentage of direct-labor cost. For many items, such as some statutory costs, there is this direct relationship. However, many of these same items do not apply beyond a certain maximum annual wage or do not apply to overtime costs, and an analysis must be made before realistic percentages can be determined. The preferred method is to list and estimate individually each item that is known to be involved in payroll burden for the project estimated. Only then can the estimator eliminate many of the inconsistencies that result from the use of an overall, average percentage. The details of a more definitive estimate are also essential for later cost control during project execution.

OTHER MISCELLANEOUS INDIRECT COSTS

In addition to engineering the FLOH costs, there are many other indirect costs, most of which can be significant cost items. Some of these indirect cost items are the following:

Freight costs (when not included with material costs)
Import duties
Taxes
Permit costs
Royalty costs
Insurance
Loss on sale of surplus materials
Owner costs

The costs of most of these items can be estimated on either a percentage-of-cost basis or in a more detailed fashion, but only after a review of historical or current data. The most important factor here, however, is to insure that all items are covered. A checklist must be developed by the CE and reviewed prior to the completion of an estimate; otherwise some item is bound to "fall in the crack."

On large projects, the net loss resulting from a sale of surplus materials can be a substantial amount of money, particularly during normal economic times or periods of high economic activity when deliveries and demand are average or long. As we have seen from our discussions in Chap. 4, the quantities and costs the CE is estimating are the averages of the actual *installed* quantities. However, to insure an adequate supply of material, the contractor will buy more than will finally be installed. Further, both design and field changes will occur on the project and these will result in surpluses. The amount of the surplus can be estimated either from an overall, historical percentage of material cost or individual percentages applied to each of the bulk material areas where the surpluses are most apt to occur (i.e., piping, electrical wiring and conduit, etc.).

Owner costs can also be a major cost item and must be reviewed thoroughly by the CE. These include such items as land costs, organization costs, and start-up costs. Here, again, checklists based on previous projects and company experience can be useful to the CE.

INDIRECT COSTS FOR SMALL vs. LARGE CONTRACTORS

There can be significant differences in the level of indirect (FLOH) costs for a small, local contractor when compared to the costs of a larger, national contractor for the same size job. There are several reasons for this:

1. The local contractors tend to maintain a permanent nucleus of both supervision and direct labor. These people are more familiar with the contractor's line of work and procedures. Also, they are more dedicated and the net result is a more productive operation.
2. The local contractor has developed a close working relationship with the local unions, is familiar with the local requirements, can avoid omissions and duplications, and, in general, can operate more efficiently with them.

Table 3 Potential Savings with "Local" Contractor

Area	Savings, %
Direct labor (higher productivity)	10
Temporary construction and consumables	50
Field supervision	40
Construction equipment/tools	35
Payroll burden	10

3. The local contractor thinks "small," whereas the national contractor is accustomed to thinking "big." For example, the local contractor will use a trailer for an office but the national contractor will revert to a larger, more permanent-type building. The local doubles up on supervision, with the site superintendent also doing the paperwork and some of the direct field supervision. The national contractor tends to organize more along structured lines, with more people (and overheads) in the field.

These factors can reduce field costs by more than 10%. Table 3 illustrates the potential magnitude of the reductions in each of the field cost areas, including a possible reduction of 10% in direct-labor cost because of higher productivity.

SUMMARY

In this chapter we discussed a number of points with regard to indirect costs that should be reemphasized:

Indirect-cost items are significant both in their cost impact and their role in cost control. They usually will warrant more than an overall percentage approach.

Design and drafting manhours/cost is the key element to determining contractor's detailed-engineering costs.

FLOH costs should be based on a prediction of the construction plan that will be used on the project.

Critical factors in determining FLOH costs are the size, length, and type of construction job and local construction practices.

Miscellaneous other indirect costs can be significant and checklists should be used to avoid omitting any of them.

Small local contractors, on the average, sustain significantly lower FLOH costs than their larger national contractor counterparts and are more efficient overall.

7

DATA COLLECTION AND MANAGEMENT

DATA vs. METHODS

In Chap. 4 we reviewed how the first of the two essential ingredients for an estimate, quantity prediction, was accomplished by using estimating methods. Now we want to spend some time looking at the second essential element, unit pricing or data. The product of these two (i.e., quantities times unit prices), constitutes the "costs" that make up our estimate. We call the gathering, analyzing, and application of these estimating data data management.

COST DATA: WHAT ARE OUR NEEDS?

As we learned in Chap. 4, estimating methods produce quantities of materials such as the weight of steel in a vessel, the meters or tons of pipe required, and the cubic meters of concrete. In addition, the methods for estimating labor requirements give us the "standard" manhours needed or, in the case of detailed-engineering work, the standard technical engineering manhours required. What we need now is the ability to convert these quantities to hardware or these standard manhours into costs. As was the case with estimating methods, it is important to realize that the data collected must be representative of historical data from the company's own experience (i.e., reflecting the way that company buys material or lets subcontracts). These data will be used in estimates that are made to predict what that company will do, which is not necessarily what the average published data reflects. For the material items we require unit prices (i.e., cost per ton, per cubic meter, per meter, etc.). For the manhours, we need first to establish the productivity relative to the standard data base to arrive at the predicted manhours for the specific location and the specific job conditions.

This subject was covered in Chap. 5. Having arrived at the manhours required, we then need to convert these to costs by using data on hourly rates. The above applies to both engineering and field craft manhours.

WHY HAVE A DATA SYSTEM?

There are a number of important reasons why serious consideration should be given to the development and maintenance of a data collection or data management system. First of all, data affect 50% of the costs in an estimate. Good methods which produce accurate quantities are ineffective in producing reliable estimates if good data to convert these quantities to costs is lacking. Securing good current data unfortunately is insufficient since this is only the first step in establishing a good system. Most data are dynamic and subject to almost instant obsolescence and a data management system is required to keep data current. A data system is needed to tailor the data to one's needs and have available at all times to the estimator current, consistent, and reliable information to allow him to prepare accurate estimates.

SCOPE OF THE COST DATA APPLICATION

The estimating methods for virtually all of the components of an estimate produce quantities in one fashion or another. To determine costs, data are needed for these areas. This includes direct material, construction labor, contractor engineering and services, and other indirect costs. For direct material, we need to distinguish between engineered equipment (i.e., vessels, pumps, furnaces, etc.) and bulk materials (i.e., piping, concrete, wiring, structural steel, etc.). We also have a third category of "delivered and erected" (D & E) equipment, such as large storage tanks or field-fabricated vessels, involving both material and labor as a single lump sum subcontract cost.

Construction labor cost data fall into two categories, "direct hired" labor and "subcontract" labor. Direct hired labor is hired and paid directly on a per hour basis to the worker by the main or prime contractor. Consequently, the data required involve wages, payroll burden, and productivity. Subcontractor labor works for a subcontractor of the main contractor and costs are normally quoted on a fixed-unit price or lump sum, often including material costs. Contractor engineering and services involves not only wages, burdens, and productivity data but also data on "overhead" costs, representing the contractors costs of running their business (i.e., rent, utilities, accounting, clerical, sales, administrative overheads, profit, etc.). Other indirect costs involve those discussed in Chap. 6 (i.e., FLOH, fees, owner costs, etc.).

PRICE BEHAVIOR

Before one can set up an effective data system, a basic knowledge of material price behavior is essential. We mentioned earlier that there are three categories of material cost items. Let us examine each of these from a price behavior standpoint.

Engineered Equipment

Engineered equipment, such as a pump or drum, are specified in detail with regard to size and design conditions and consequently are tailor-made for the specific project. These items are made in fabricating shops by vendors who specialize in making that particular line of equipment. Therefore, shop load is sensitive to demand and, living in a competitive environment, vendors are forced to adjust their prices accordingly. Because demand varies substantially depending on local economic activity and on the vendor's ability to keep their shop occupied, there can be significant price fluctuation in this area.

Estimators must be capable of predicting these price fluctuations in advance and including the impact in their estimate, either in the form of increased costs for specfic equipment items or as a general "activity" allowance. The problems associated with analyzing and predicting economic activity and its impact on costs are discussed in more detail in Chap. 9.

Bulk Materials

Bulk materials items include piping, concrete, steel, and electrical wiring, and represent the support material items for the engineered equipment. Since they are required in large quantities (hence the term "bulk") and since each item is not engineered especially for the project, they can be manufactured in quantities beforehand and sold more or less "off the shelf." Because they are manufactured for stock and have a broader market than engineered equipment, their price behavior tends to be more stable.

Delivered and Erected Equipment

As we mentioned earlier, delivered and erected equipment consists of engineered equipment with a significant amount of construction labor involved. The material portion is subject to the same fluctuations in demand and hence price as engineered equipment. The labor portion tends to be more stable and its behavior is a function of local field conditions.

As a result of price behavior, material-cost data are dynamic and are subject to instant obsolescence. The CE is faced with the problem of setting up a data management system that reflects these changes, not only as they occur but before they occur, since an estimate is always a prediction of the future. The task is further complicated by the massive amount of the data that must be accumulated and managed, especially if the CE works for an international company and requires the ability to predict costs all over the world. Table 1 is an example of the large amount of data which needs to be collected, analyzed, and maintained for chrome alloy piping. Similar needs exist for other piping materials.

If we consider all the possible items that could be installed in a typical process plant, we need to manage in the range of 150,000 to 200,000 pieces of data. This then has to be multiplied by the number of locations involved in the data system. Consequently, it is not unusual to manage literally millions of pieces of data. Obviously, to collect, publish, and maintain fully detailed estimating data for all locations would be a nearly impossible and inefficient task, considering the time and manpower involved. To see how this can be more reasonably accomplished, let us examine two different approaches to the problem.

TWO APPROACHES TO DATA COLLECTION

As was the case in estimating methods, there are two basic approaches to data collection. The one approach is to gather specific data for a specific estimate as you are doing the estimate. This is normally accomplished by contacting vendors for preliminary quotations for each area of the estimate (the follow-up to arriving at the quantity requirements for a definitive estimate by carrying out preliminary engineering work). This more detailed approach is normally used by a contractor in preparing estimates and was discussed in Chap. 4. This approach

Table 1   Chrome Piping Materials Data

Valves

    Standard prices in each of 23 sizes — ½ in. — 36 in.
    7 flange ratings
    4 kinds of valves

In addition

    2 types flanges in 7 flange ratings
    2 types fittings in 7 schedule numbers
    Pipe in 7 schedule numbers
    Unit prices for various shop fabrication tasks
    Etc.

Altogether 1500—2000 "Standard" costs for each chrome alloy

    5000 Prices including 1-¼%, 2-¼%, 5% chrome

requires little in the way of formal data collection or management since the data is "re-collected" for each job.

It is the "in-house" data system that requires an extensive amount of management because of its complexity and volatile nature. This is especially true if the CE requires data for more than one location or one country. One way of solving this problem is to set up a location index system to provide a practical way of estimating material prices at a specific location and a specific point in time. With this system, detailed costs for all of the material components going into an estimate are established for a single well-defined location time and set of conditions. These are what we will call "standard" or "base" costs. These standard costs are then converted to the individual locations by means of indexes. These indexes express the relationship of costs at the particular location and time to the standard or base costs. With a system such as this, feedback from actual experience on individual projects can be used effectively to update and improve the data. This feedback closes the loop and provides the CE with current information needed for producing estimates.

STEPS TO SETTING UP A DATA SYSTEM

Setting up a data system within a company for use on all future estimates is not a difficult or time consuming task. The system can initially be simple and gradually expanded to handle all of the company's cost engineering requirements. The important consideration is to start the system so that you can reap the many benefits of a formal, structured system rather than a haphazard, "bottom drawer" approach, which usually results in inconsistent and unreliable estimates. The simple steps to establishing a data system are the following:

Establish the specific data needs
Gather all available data, both historical and current
Analyze the data, condition and normalize (remove abnormalities)
Identify time frames
Establish a single source (i.e., a "data book")
Set up a feedback and maintenance system
Use factors to keep current

Let us take pipe material costs as an example. The cost of any given piping material (i.e., carbon steel, stainless steel, alloy, etc.) can be established rather easily from a manufacturer's price list. This price list can be the first page of your new data book. Even if the price list is out of date, it will serve the purpose, since the most important consideration is the price relationship between different pipe sizes. Then, as feedback information on actual prices is received from purchase orders, these can be compared to the "standard" costs and factors developed. These factors would be used in estimates prepared from

Table 2 Factor Example

| | | Standard prices valves | | | | | At location/time prices | | | | |
| | | Size | | | | | Size | | | | |
Flange rating		2 in.	4 in.	6 in.	8 in.	10 in.	2 in.	4 in.	6 in.	8 in.	10 in.
150		10	20	30	40	50	8.50				
300		15	30	45	60	75		24.00	23.60	49.00	
400		20	40	60	80	100			64.10		
600		40	80	120	160	200					82.00

Factor Calculation

8.50/10 = 0.85   49.00/60 = 0.82
23.60/30 = 0.78   82.00/100 = 0.82   Average = 0.812
24.00/30 = 0.80   64.10/80 = 0.80

the standard data book costs. Table 2 is an example of the development of factors to be applied to standard valve prices. It should be noted that the factor developed in the example represents the cost at the particular location where the purchases were made and for that particular time period. The setting up of these factors, or location indexes as they are often called, would need to be developed for each location where the company is likely to build or for which estimates will be required. The development of these location indexes are discussed below.

CONVERSION TO OTHER LOCATIONS

The mechanism for converting standard costs to present-day costs at a given location is accomplished by means of cost indexes. These indexes are factors representing a comparision or cost ratio between the latest cost data available and the equivalent standard cost data. The CE has analyzed the feedback data (usually in the form of purchase orders from recent projects) and arrived at an average ratio, which is called a location cost index, as illustrated in Table 1. Table 3 is an example of the cost indexes for various locations for the pipe example discussed earlier. Several important points should be noted from this table and our previous discussion:

By using the indexes, the standard or base costs never have to be changed; only the indexes change.

The indexes can be updated as often as desired. Because of the volatile nature of the material costs in most locations, quarterly review and reissue of the indexes are necessary.

The indexes must be tied to a time frame, preferably as close to present day or the time of estimate preparation as possible (i.e., for Table 1, 2Q83). Usually, because of the time required to analyze and publish the data, there is at least a time lag of one quarter.

The same approach conceptually followed for materials can be used for developing indexes for engineering, construction labor, and indirect costs. The final step of converting the present-day costs to the time when the material will

Table 3 Alloy (Chrome) Piping Cost Indexes, 2Q83

Place	Pipe	Fittings	Flanges	Valves	Etc.
New Jersey	1.56	1.35	1.42	1.39	
Italy	0.74	0.65	0.51	0.46	
Japan	0.77	0.55	1.27	0.83	
Etc.					

actually be purchased involves adding predicted escalation for that time frame. How this prediction is developed and made is discussed in Chap. 9.

SOURCES OF COST DATA

We mentioned briefly that a common source of material cost data that is needed to develop the cost indexes or factors is purchase orders on current or recent projects. This is not the only source. Some examples of cost data sources are shown in Table 4. Although the sources for data appear to be numerous, in actual fact the data are not readily available and are difficult to analyze (or measure, in the case of productivity). In the case of survey and personal contacts, the results are often laced with subjectivity.

Cost surveys, done properly, however, can be an excellent source of data. Cost surveys usually involve three phases. The first phase consists of preparing a comprehensive questionnaire which is sent to the prospective contacts at the location surveyed. These contacts would include affiliate companies, contractors, subcontractors, vendors, engineering companies, consultants, and other owners. The second phase covers the visits when the contacts' answers to the questionnaire are discussed first hand, in detail. Finally, the last phase involves the analysis of the results and the development of new data—both current unit price data and future escalation predictions, usually based on an analysis of the responses with regard to future economic activity. Often, an overall summary of the results, suitably disguised as to individual source, is sent to each of the contributors as an incentive for cooperation on future surveys. Surveys can be

Table 4   Sources of Cost Data

Feedback from projects

     Purchase orders
     Engineering contracts
     Wage rates
     Engineering manhours/productivity
     Labor manhours/productivity
     Etc.

Literature

     Standard building materials
     Wage rates
Manufacturers' published price lists
Union agreements—wage rates and benefits
Vendor contacts
Affiliate contacts and feedback
Published indexes
Surveys

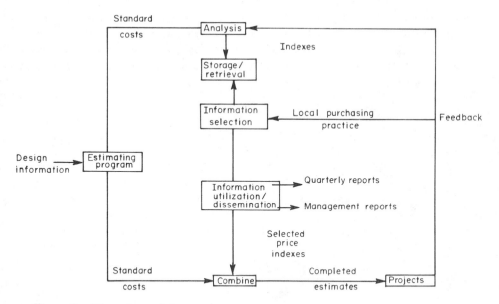

Figure 1 Flow chart of data management system.

periodic for locations where continuous capital investments are made or considered or specific surveys can be organized and carried out for new locations.

SUMMARY

Figure 1 summarizes in flow plan form the procedure followed in gathering, analyzing, publishing, and updating data (i.e., data management). The standard cost data are a one-time effort and remain fixed, whereas the right-hand loop is a continuous one, starting with analysis through to feedback and then starting with analysis all over again; the end result is updated price indexes in the form of quarterly reports for use in current estimates. Because of the immense amount of data involved, a computer is essential to the system. In summary, in this chapter we learned that data management—like its counterpart, estimating methods—is an essential ingredient in the preparation of an estimate. In fact, the results of methods, quantities, multiplied by the end product of data management, current unit prices, produce the total estimate. Current unit prices can be established in-house by setting up standard costs and multiplying these by cost indexes. The cost indexes are factors or cost ratios between actual current costs and the standard costs. We also learned that the in-house approach can involve an immense amount of data collection and considerable effort but can be started in a simple way and gradually expanded. Maintenance of the system is by far the more difficult and time-consuming step. The definitive or contractor approach also requires considerable effort but is more in the direction of establishing specific rather than average unit costs by obtaining quotes (as discussed in Chap. 4).

8

ESTIMATE CODING AND DOCUMENTATION

GENERAL

The previous chapters discussed the basic principles involved in cost estimating and many of the techniques involved in developing the capacity to prepare accurate and reliable estimates. One area not yet discussed that is often overlooked despite its importance is the documentation of the estimate basis and details. You will recall the importance we placed on consistency, clarity, simplicity, and accuracy when we discussed such areas as estimating methods. Those same critieria must be applied to the estimate preparation; otherwise we will not achieve our original end objectives and all the effort that has gone into the development of estimating methods and data management will have been to a large degree wasted. In this chapter we will discuss estimate coding and estimate documentation, the two essential elements needed to insure that our final product, the estimate, can be fully used in the manner we have been discussing. Because this area is accounting and paper-oriented rather than technical, it does not receive the emphasis it should. As we will recognize as we go through the chapter, there are many valid reasons for giving this area as much attention and effort as the more technically oriented ones. Fortunately, not much additional effort is needed to meet the minimum requirements for good estimate coding and documentaion.

ESTIMATE CODING

Throughout the previous chapters we discussed a multitude of estimate terms and, in particular, categories of work that are involved in building a project. For each of these we need to prepare estimates in terms of materials, labor, engineering, FLOH, and other indirect-cost items. It makes sense, with the large number of items and categories involved, that to maintain some kind of order it is necessary to segregate these costs. This is not only true during the estimate preparation stage but is essential throughout the execution cycle. As we will learn in later chapters, contractors will want to segregate costs to keep track of where and how they are spending their money (i.e., purchasing of materials, expenditure of engineering and field manhours, etc.). They will want to compare these costs with costs in the control estimate. Also, CEs want the actual cost data to be segregated when they feed this information back to the home office for analysis and data management. This segregation, as we have been calling it, is referred to as a "code of accounts" and, more specifically, when it is used during project execution, it is called a "construction code of accounts."

WHAT IS CODING?

Put simply, codes are the umbilical cord between cost accounting and cost engineering (cost estimating and cost control), since in most cases the estimate could not survive as a control tool without data supplied by accounting. The sole mechanism for receiving these data is the construction code of accounts. This code is the dictionary for the language used for all the cost records and, as such, is an essential ingredient for cost forecasting and cost reporting, two main elements of cost control, as we will discover in later chapters.

Figure 1 is a sample code breakdown for a typical process unit. As you can see from the table, codes are broken down into various cost levels, each level progressively more detailed. The primary code level represents the cost of an entire process unit where there is more than one unit involved in the project or the code can also be used to accumulate costs for large geographic areas, such as on-sites, off-sites, a specific tank field, etc. There are few primary codes for a project and they would be identified simply as "pipestill" or "off-sites." The primary code is the first one or two numbers in a code and immediately tells the reader to what unit or area the item (represented by the numbers or codes that follow) belongs. This code is referred to as "area" code in the example in Fig. 1.

The secondary or main code in our example is the one that immediately follows the primary code and indicates the major cost category (i.e., is it a material or labor or engineering item?). The tertiary code and the codes that follow (subcodes and detailed codes) are the heart of the code of accounts. The first two digits can be allocated to a specific material category, such as piping. The third digit can then be piping valves or piping flanges. Depending on the amount of detail required and the capacity of the accounting program, additional "drop numbers" or digits can be used to get further levels of details, such as the materials of construction (i.e., carbon steel, stainless or alloy materials) or type (i.e., gate valves, check, globe or ball valves, etc.).

Area		Main		Sub		Detail	
Name	No.	Name	No.	Name	No.	Name	No.
On-sites	01	Labor	1	Excavation	0	Hand excav.	001
Off-sites	02	Material	2	Concrete	1	Mach. excav.	002
Utilities	03	Subcontract	3	Str. steel	2	Trenching	003
				Equip. and mach.	3	Backfill	004
				Electrical	4		
				Instruments	5	Formwork	100
				Piping	6	Rebar	101
				Paint and insul.	7	Pour found.	102
				Buildings	8	Pour slab	103
				Indirect	9		

Example:

01	—	1	—	1	—	100
Onsite		Labor		Concrete		Formwork

or labor for setting on-site formwork

Figure 1 Sample code breakdown for a typical process unit.

WHAT IS NEEDED FOR A GOOD CODING SYSTEM?

Table 1 lists the five important criteria that must be met if we are to have an effective coding system. First, the code must be a reflection of the way the job will be executed and the way costs can reasonably be collected. Most company standard construction codes of accounts developed only after studying in detail how the company executes a job and how it collects the costs. In most cases, execution follows a similar sequence. As a result, even though most contractors have their code of accounts, there is little difference between the various code systems used. For example, all contractors have a separate code for piping, broken down pretty much as we discussed earlier. They also all would probably have a separate code for shop-fabricated or field-fabricated pipework and another for shop vs. field fabrication of a vessel. The important point is that, like the estimating methods, the code should reflect the way the particular contractor will carry out job execution. Thus, the estimate (which predicts job execution) is made to match the code.

Secondly, the code should be detailed. As we will learn later, details are essential to cost control. CEs cannot analyze an overrun in piping if the feedback from the monthly cost report only shows a total piping cost (one secondary code only, with no drop codes). They need more detail and a good code system will provide this detail. One word of caution, however: detail for detail's sake should be avoided. For example, if concrete foundation work is normally collected and monitored in the field on a manhours expended per foundation or per cubic meter basis, there is no incentive to further subdivide the code for this work into excavation, formwork, bar reinforcing, and concrete pouring.

The codes adopted by a company should be documented in detail. A code book should be published and made available to all key personnel involved. The book should contain a code by code listing, with a detailed word description of each code. Where there are "gray" areas or areas of potential overlap or confusion, the codes in question should not only have their desciptions listed, but also what is excluded (i.e., the items that could easily be misunderstood, as is often the case with, say, instrument wiring: is it in the instrument code or the electrical code for that particular company?). Finally, the construction code of

Table 1   Criteria for an Effective Coding System

Reflect job execution approach
Detailed
Document description
Both inclusions *and* exclusions
Geared for project cost control

accounts should be geared for project cost control only. Although some owners have a construction code of accounts of their own, many do not and rely on the contractor to supply one. [Even when owners do have their own, it is probably more efficient to allow contractors to use a code they are thoroughly familiar with (i.e., their own.)] In any case, all owners have a code of accounts that they use in running their business, which is normally not the execution of capital projects. The owner's codes are set primarily by their internal business requirements (i.e., manufacturing, production, etc.) and by local and national tax requirements. Consequently, they are of little use in project execution and should not be imposed on the project contractor. If owners need for depreciation and/or tax purposes a cost breakdown on their code basis, this can be provided at the end of the job by the contractor by rearranging or recasting their construction costs into the owner's code. Since the contractor's code should be in considerable detail for cost control needs, there will be adequate information available to do this. If, for some reason, cost breakdowns in the owner's code are still required on a monthly basis during project execution, often an estimated breakdown, using "engineering judgement," is acceptable or, alternatively, the accounting computer program can be reprogrammed to allow a separate "dump" in a second code system by cross-referencing codes. In any event, under no circumstances should the contractor be forced to use a nonconstruction code of accounts during project execution.

WHY DOCUMENT ESTIMATES?

Preparing and releasing an estimate is similar to publishing a textbook. The estimate will be used ("read") over and over again by both the owner's and contractor's personnel for several years after it is released. It will be revised from time to time (project changes) and experience in using the estimate will be fed back to the home office for use in future estimates. As we will learn in later chapters, in using the estimate as a control tool for cost control, detail and clarity of details is essential. The cost control engineer needs to know why and how cost estimators predicted something was going to happen if they are going to properly make a cost analysis. Also, once the estimate is released to the contractor, reallocations or recasting is necessary to put the estimate into the contractor's code of accounts or to reflect a different execution approach. Further, changes, unfortunately, seem to start from day one and we need to know from what basis the change is made. In other words, we must have a document showing us in a clear, precise, and comprehensive manner the basis for the estimate and a clear presentation of the details.

HOW CAN WE DOCUMENT ESTIMATES?

As is often the case, it is the simple and easy steps that can accomplish most of what we are looking for in estimate documentation. Table 2 lists a half-dozen

Table 2   How to Document Estimates

Standardize forms
Write down all calculations/assumptions
Date all calculation sheets
Document basis, concise and clearly
Emphasize neatness
Maximize computer usuage

steps that, if followed with proper discipline, can accomplish our objectives. Using standard forms not only provides consistent-looking detail sheets but also disciplines the cost estimators. They tell them where to put the data and can inspire in them some degree of neatness. Different forms can be designed for various parts of the estimate, always with a view to consistency, neatness, and the saving of time (i.e., preprinting of headings, columns, formulas, etc.) Figure 2 is a sample of an estimating calculation sheet. Estimators should be encouraged to write down all calculations and assumptions made in the course of preparing the estimate. Too often, a critical calculation is done on a piece of scrap paper and thrown away, leaving the poor cost control engineer, a year later, in a position of trying to analyze how and why a particular cost was calculated. Assumptions are even more critical, since the very basis for the estimate is the predictions (assumptions) made by estimators. This information is essential for later cost analysis (comparing actual vs. predicted performance). The simple step of dating and initialling all calculation sheets and forms can help eliminate later headaches when CEs and managers try to reconstruct what happened. This is especially applicable where many screening estimates are made or there is recycle on budget estimates, and reconciliations are required between the various estimates (i.e., who did what, when and why?). Formally documenting the overall basis for the estimate is extremely important. A project is a dynamic, moving operation and so are the costs associated with the project. Therefore, it is important to document what the basis for the estimate was at the cut-off time (i.e., the situation on the project at the time the estimate was made). Table 3 lists some of the more important estimate basis information that should be included in the estimate documentation.

The estimate documentation can be done in two stages. An estimate basis memorandum can be prepared before the estimate is started, outlining the intended basis and the procedure to be used in preparing the estimate. This is particularly helpful on large projects where many estimators and project people are involved and need to be brought on board on a consistent basis. The second stage of formal documentation would occur after the estimate is completed, when a package would be put together, often called the "scope book." This book documents the scope of the estimate and its basis. The specifications,

Code	Description	No. units	Material cost			Labor cost			Total M + L ($)
			Source	Unit cost	Total ($)	Manhours/ unit	Manhours ($)	Thousands Total ($)	

By _____ Estimate _____

Date _____ Location _____

Figure 2 Sample of an estimating calculation sheet.

Table 3   Estimate Basis Information for Documentation

Scope basis at cut-off date
Assumed project schedule
Source for material purchases
Detailed design basis (local design codes, etc.)
Escalation rates assumed
Contracting approach
Labor source/rates/productivity
Contingency basis

including revisions up to the date of the estimate, would be included along with all appropriate drawings. Also included in the overall package would, of course, be the estimate details prepared on standard forms in a neat, comprehensive way, as described earlier.

In discussing documentation, we cannot overemphasize the importance of neatness. Unfortunately, this trait is not all that common among engineers and managers, particularly when it comes to filling out forms and documentation of calculations. Consequently, they have to be disciplined by some degree of fomality (i.e., forms, instructions, etc.). Fortunately, more and more of the calculations and the documentation is done by the neatest engineer and manager of them all, the computer.

SUMMARY

In this chapter we have discussed what too often is considered a minor and insignificant area of cost engineering: cost coding and estimate documentation. We have seen that this area is important. In many ways coding and proper estimate documentation are vital to producing and using effective estimates. Some of the major points discussed follow:

Codes are the umbilical cord between cost accounting and cost engineering. There are two basic types of codes of account, related to two basic needs: that of control of costs during project execution (a construction code of accounts) and the codes required by an owner to carry out day-to-day business. The CE should avoid situations where the owner code is used for project execution.

Codes can be developed to provide virtually an infinite amount of detail by applying drop numbers to the basic code number. However, detail for detail's sake should be avoided.

Construction codes of account reflect the way a job will be executed, hence most construction codes of accounts are similar.

Estimates will be used over and over again after they are first issued. It is essential that they be documented in a clear, concise manner.

Documentation requirements shown in Table 3 should be followed religiously. They are simple requirements but the return on the small amount of effort is many-fold.

9

SOME PROBLEM AREAS IN COST ESTIMATING

GENERAL

A cost estimators primary function, the responsibility of predicting the future, is not all peaches and cream. Just as a stock market analyst or an economist has problems in trying to establish what the future holds, so the CE runs into more than his share of problems when he takes on the fuzzy or gray areas of a cost estimate. In previous chapters, we discussed two common problems: the estimating of piping costs and the predicting of field labor productivity on a specific project. You will recall that we advocated tackling both problems in a logical, "scientific" way, considering all factors present that can effect the final execution and hence the estimate prediction. Experience shows that this approach is far superior and more rational than one that involves the use of overall, gross percentage "allowances," an approach that is based more or less on the philosophy that it is impossible to either estimate or control these two areas.

There are several other areas that are also critical and, unfortunately, even fuzzier than piping and labor productivity (i.e., the estimating of escalation and contingency and the overall problem of estimating a project that would not follow an average historical pattern—what we call a "revamp" project). This chapter will be devoted to discussing the first two estimating problem areas and a recommended approach to predicting their cost impact on any given project. Because each of these are major subjects, we will cover the subjects lightly, concentrating on principles rather than specific details. Chapter 10 will be devoted to the problems associated with estimating a revamp project.

ESCALATION

In Chap. 8 we discussed the management of data and the use of indexes, which allowed the CE to prepare an estimate of present-day costs (i.e., the cost of the project at the time the estimate was prepared). For this estimate to be valid, the project would have to be built in that immediate time frame (i.e., engineering, procurement, and construction would all have to happen right then, to match the predicted present-day costs used). This, of course, is impossible and could never be the correct prediction. The basic design work has probably just been completed, and 2 to 3 years of detailed design, procurement, and construction (i.e., project execution) lies ahead. Therefore, to arrive at the estimate of the final costs, the cost estimator must add the increase (or decrease) in costs resulting from the impact of escalation. This is done by multiplying the annual escalation rate by the escalation time period to arrive at the total escalation rate. This rate is applied to the present-day costs to arrive at a total escalation cost. The escalation cost then is added to the present-day cost to arrive at the predicted final cost.

ESCALATION TIME PERIODS

Figure 1 illustrates a typical overall project schedule in bar chart form and the recommended average escalation time periods for each of the major components

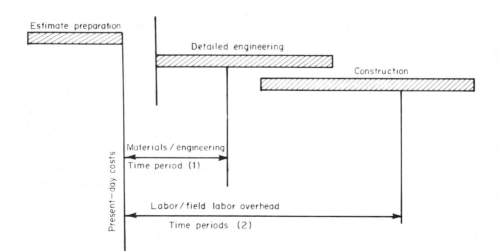

Figure 1   Escalation time period. (1) Normally contract award plus about 6 months; (2) normally at 70% of construction time.

of an estimate. Note that each time period is measured from one common starting point, the time of the estimate preparation or, as we have been referring to it, present day. All times have been taken to the "centroid" or weighted-average-time period (i.e., material to the time when half the value of all purchase orders have been placed and half are yet to go). The same holds true for engineering, field labor, and FLOH. In the case of field costs, an average of 70% of the construction time transpires before half of the manhour value is spent. This is caused by the slow manpower buildup in the early months before peak manpower is reached.

MATERIAL ESCALATION

Since most cost data relating to material costs in general and equipment costs in particular have recent purchase orders as their source, the present-day costs included in the estimate represent the cost of a purchase order if it were placed at the time the estimate was prepared. Except in times of extremely high economic activity (i.e., when business is really booming at an abnormally high rate), most purchase orders are quoted at a fixed or lump sum cost. Consequently, the vendor has predicted total cost, including any escalation that may occur from the purchase order date until the piece of equipment is fabricated and delivered to the job site, a period of 6 months to over a year. However, the purchase order won't be issued present day and the estimator needs to add the additional amount of escalation for the period from present day until the purchase order is actually placed, but no more, since the vendor will include escalation beyond that point.

SELLING PRICE vs. COST

Figure 2 is a plot of escalation in the United States over a 15-year period, 1966-1981. The graph illustrates a number of important fundamentals that must be understood if escalation is to be rationally predicted. Note first that escalation is cyclic, following what we call the "activity cycle," (i.e., as business activity picks up the escalation rate increases; as business falls off, the escalation rate moderates and can even become negative for a period).

Second, note that despite these cycles, there is a long-range trend; unfortunately, constantly upward. Until the sharp break in 1974-1975 (as a result of both high activity and the oil crisis), this trend line had averaged around 5% a year (overall escalation) for the United States and most parts of the world. Note that there is a constant driving force from the peaks and valleys of the cycles back to the level of the long-range trend line. Therefore, the historical data and escalation behavior illustrated in Fig. 2 can be useful to the estimator in predicting future escalation. They also make the estimator aware of a basic

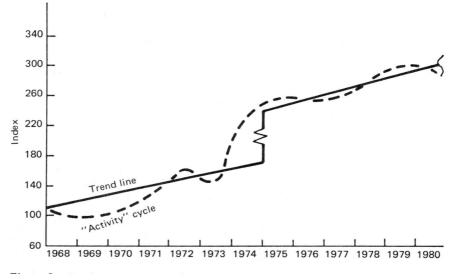

Figure 2   Escalation, short- vs. long-range.

fundamental that must be considered when trying to predict future costs, namely, that the selling price of an item is determined more by the level of economic activity (i.e., competition) than by the actual costs of making the item. For example, during the 1974-1975 period of high activity, the costs of materials went up at an annual escalation rate of 50% or, in some cases, even more. Yet the items of costs that made up the actual cost to the vendor (i.e., raw material, such as steel plate, castings, etc.; shop labor wage rates and shop overheads, engineering costs, etc.) went up at a rate only slightly above normal (say, 10 to 15% annual rate). The difference is a result of higher demand than capacity. Therefore, predicted escalation rates for materials and subcontracted costs should never be based on a synthetic buildup of predicted escalation rates for raw materials and labor but rather on a more detailed study of Fig. 2.

SOURCES OF DATA FOR ESCALATION RATES

Data similar to those shown on Fig. 2 are readily available from publications and periodicals, and are useful for establishing overall escalation rates.

A word of caution, however—any overall escalation rate is a function of the commodity mix used to establish the rate. Estimators need to analyze the mix for any published rate to insure that it is representative of the type of project and industry for which they are preparing their estimate. For example, many government escalation indexes or rates will include a broad commodity mix, including such items as glass, leather and paper, items which are not representative of the material items used in the design and construction of a typical process plant.

Estimators, however, require more specific rates for each estimate category if they are to prepare a definitive or detailed estimate. In other words, they need, at minimum, annual escalation rates for overall material, engineering, labor, and FLOH. More comprehensive estimating systems would require rates for specific material items and specific labor disciplines. Most of these specific data are not readily available. Labor rates can easily be established from union agreements, which normally span a 1- or 2-year period and are predictable beyond the expiration date. Material rates, however, require more effort and here firsthand surveys must be made to develop the best estimate of future economic activity in addition to increases in raw material and labor costs. This is done by interviewing contractors, vendors, company affiliates, and analyzing such data as engineering backlog, shop orders backlog, status of shop capacities, delivery terms, and announced future projects. In all cases, CEs need to check their final conclusions against Fig. 2 or similar data to insure that they are consistent with the long-term picture.

LONG-RANGE ESCALATION

Occasionally CEs are faced with preparing estimates that try to predict the cost of future projects, not to be built for another 3 to 5 years. This is particularly true of screening estimates for new ventures or new market areas. In these cases, estimators have no alternative but to make predictions using present estimating methods. These methods are based on present technology, which can and does change. There are data to show that, in some areas, the effects of escalation can be offset to a large degree by the credits that accrue from new and improved technology. W.L. Nelson, a recognized authority on indexes and escalation in the process plants industry, completed a study showing that the average per barrel cost of an oil refinery actually did not change over a 20-year period because the increases caused by inflation were offset by both the advances of technology and the economics of size (i.e., current refineries are much larger and, hence, less costly per barrel or thoughout). Later studies of the unprecedented escalation spiral in the 1973-1975 period show that this conclusion does not apply during these unusual periods. However, under more normal circumstances, the impact of technology can be significant and CEs must consider whether this will apply to any estimates they prepare for future projects.

ESCALATION EXAMPLE

Table 1 is a simple example of how escalation is estimated for a total project estimate. There are a number of important points in the example which are worthy of note. The escalation time base represents the time frame for the latest cost data from the data book (i.e., the latest quarterly report on cost indexes or factors available to estimators in their data book). Therefore this is their "present-day" time, even though there is a difference (lag) of a year between the "present-day" and the date the estimate was completed. This time must be added as escalation time, as indicated in the example. Also added is the time between estimate completion and CA or contract award. Both of these time frames will vary from project to project and must be established by estimators prior to calculating the cost of escalation. The example shows escalation calculated as a separate identifiable cost for each major category. This is an acceptable approach for early, order-of-magnitude estimates but is not recommended for more detailed estimates. In the latter case, escalation should be added to each individual item in the estimate, so that a total estimated final cost is presented for each item. In this way, an "apple to apple" comparison can be made between actual and predicted costs during project execution. As will be seen in Part II of the text, this is essential for cost analysis and effective cost control.

CONTINGENCY: WHAT IS IT?

One of the more controversial and least understood items in every estimate is contingency. Contingency can mean different things to different people and therein lies the estimator's problem—primarily communications. Each manager, each engineer and, often, each estimator has their own definition for contingency, and although they will likely differ, they could all be correct, provided that the definition they are using is a consistent and well-defined approach and is recognized and understood throughout their company.

There are essentially two components in any estimate—a base estimate representing the known and firm items in the estimate basis and a provision for uncertainties. The base estimate includes all those items which are both defined and most likely to happen. Usually this information is in the form of a job specification and estimate basis memo. The cost engineer prepares the estimate on the most probable costs for each of these known items (i.e., the items that will most probably be required or the event that will most likely occur and the estimated costs represent the most likely final cost). An estimate prepared in this manner is therefore unbiased and should have a 50/50 chance of overrunning or underrunning. Unfortunately, this is not always the case and there are a number of possible reasons why the estimate may need to be adjusted to bring it to this 50/50 "most probable" level. Most of these reasons are associated with uncertainties in the estimate basis or in the estimating methods and data. Table 2

Table 1 Escalation Example

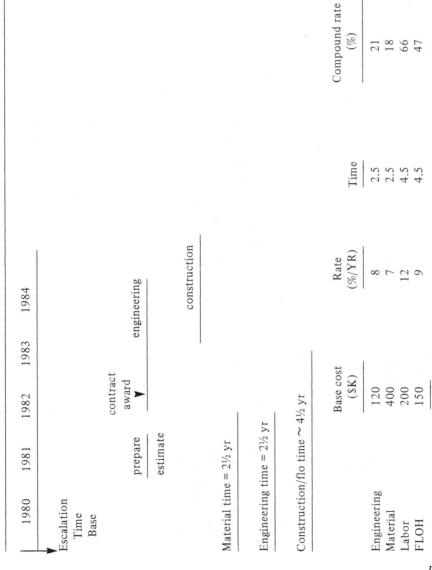

	Base cost ($K)	Rate (%/YR)	Time	Compound rate (%)	Escalation ($K)
Engineering	120	8	2.5	21	25
Material	400	7	2.5	18	72
Labor	200	12	4.5	66	132
FLOH	150	9	4.5	47	70
	$870				$299

Material time = 2½ yr

Engineering time = 2½ yr

Construction/flo time ~ 4½ yr

Note: K = thousands

Table 2   Estimate Uncertainties

Inadequacies in estimate basis definition
Inadequacies in estimating methods and data
Concerns regarding "soft" areas in the estimate
(escalation, productivity, schedule etc.)
Chances of overrun
Untried process
Extent of errors and omissions
Weather, strikes, accidents, delays, etc.
Abnormal activity

---

lists some of the more common uncertainties faced by the estimator and/or management. There are many other uncertainties, some real and some imagined or influenced by the estimator's or manager's desire to ensure that the estimate is "adequate."

Recommended Approach

One of the purposes of this section is to recommend a rational approach to the development of this provision for uncertainties and thereby produce unbiased and objective estimates which can be effectively used for decision making, funds allocation, and project planning and control. To do this, it is first necessary to separate uncertainties which are common to most projects and which can therefore be analyzed based on past history from those uncertainties which are possibilities or concerns but which may or may not happen. It is recommended that the "known" uncertainties be called "contingency," whereas the possibilities or concern adjustments be called "allowances." Let us first define and examine "contingency."

Contingency — A Recommended Definition

In our approach, contingency is defined as "a specific provision of money or time in an estimate for undefined items which statistical studies of historical data have shown will likely be required." Specifically, the contingency covers any inadequacies in estimate basis definition, both design and execution, and inadequacies in estimating methods and data. Most, if not all, estimates are subject to variation resulting from inadequacies in both areas and, therefore, a contingency will always be required.

Inadequacies in Estimating Basis Definition

Every estimate has a basis. For early estimates, much of the basis is developed from assumptions and extrapolations of past history and little is well defined and not subject to change. As the project develops through its life cycle, the additional engineering work and execution planning creates better definition, a better basis, and a more accurate estimate, as we discussed in Chap. 2. In most cases, the additional definition results in changes to the original assumed basis and, unfortunately, historically, these changes average on the plus side. These changes occur throughout the project cycle. Consequently, any estimate prepared at any time during the project life cycle will be biased low by the extent that the basis is inadequate. For early estimates, 30% to 50% may have to be added to allow for this, whereas in estimates prepared after the design and execution basis have been reviewed and developed in detail, the contingency can be as low as 3% to 5%. Since these "changes" occur on all projects, the amount required can be determined by analyzing past history. The amount required will be a reflection of the degree of definition normally supplied at any given time and the degree of control exercised on changes after a definition is agreed. Therefore, the allowances required for inadequacies in estimate basis definition is a function of how a company operates and, as a result, the amount will vary from company to company. It is important to recognize that changes we are discussing here are changes within the original scope of the project (i.e., changes resulting from development of the original design and execution basis and not major changes in scope such as increased plant design capacity, location, major schedule changes, etc.). Scope changes are discussed later under "Allowances."

Inadequacies in Estimating Methods and Data

No estimating method or cost data used by the estimator in preparing the estimate is perfect and, again, like estimate basis definition, inadequacies here historically result in an estimate biased on the low side. The allowance that needs to be added to account for this is normally in the order of 5% to 10% but will depend on the quality of the individual company's cost engineering methods and data and its past track record. Whenever possible, the amount of contingency added for this item should be kept to a minimum (5% or less) by improving the quality of the estimating methods and/or data.

CONTINGENCY FOR ESCALATION/OTHER "SOFT" AREAS

Often, a contingency or allowance is added to an estimate because the estimator or managment lacks confidence in a particular part of the estimate. These "soft" areas are places where there appears to be insufficient information or where experience has shown that there is a high degree of variability. Examples of this are the very estimating problem areas we have been discussing in this chapter and previous chapter (i.e., escalation, labor productivity, and piping). As was the case with inadequate estimating methods, this type of contingency allowance should be avoided and, instead, the soft area should be firmed up by seeking out additional information and putting oneself in a better position to predict the most probable cost. We discussed how this can be done "scientifically"when we reviewed each of the soft areas in this chapter and earlier chapters.

BIAS CONTINGENCY

Often a contingency will be added to minimize the possibility of an overrun. For example, if a normal 10% contingency when added to an estimate prepared at a certain time in the development of a project statistically results in a probability of one chance in five that the estimate will overrun by more than 10% (the normal maximum cushion given by management on an appropriation before reappropriation is needed), the local management may want to increase the odds that there will not be an undesirable overrun and a need to go back to top management for more money. They can do this by adding a "bias contingency" (i.e., artificially creating a high estimate). The amount added to get a desired probability can be determined by statistically analyzing past history on similar estimates. The addition of a bias contingency is not a recommended practice, since it creates an estimate that is not a realistic prediction of job execution. This high estimate can "kill" a project by producing overly pessimistic economics or, alternatively, if the project is approved, it creates a "fat" control estimate and, in all likelihood, the additional money will be spent. If the money is not used for unknown occurrences, probably it will be used for financing an extra, unappropriated pet project. The preferable route is to test the project economics for its sensitivity to an increase beyond the allowable overrun and then make a decision. If the decision is to approve the project, only the most probable estimated cost should be approved and used as the control estimate.

PROCESS CONTINGENCY

In cases where the project estimated is a new and untried process, the process engineers may have some genuine concerns regarding the capability of the design basis equipment to fully meet the design requirements. For example, when the plant is finally built and started up, there may be a need to add additional heat exchange capacity or pumping capacity. Because this is a new design, the designer can't be certain of the actual requirement, but history shows that abnormal additions or changes are often needed during both the detailed design and startup. To cover this uncertainty resulting from inexperience, a "process contingency" is sometimes added. The amount of this contingency is usually determined by risk analysis [i.e., the estimator and the process designer look at each of the soft areas of the design basis and the designer makes a prediction of the probability of having to modify the design later (this probability will always be less than 50% since the design basis is the most probable basis)]. He will also define for the estimator what the modification might be. Based on these definitions and probabilities, the estimator calculates what contingency must be added to cover these possibilities (usually the product of the probability and the estimated cost of the modifications). Since most processes and projects have previous experience behind them, this type of contingency is uncommon. However, if used, it is extremely important that the estimator document the basis for the process contingency and separate this item from the normal project contingency described earlier.

ALLOWANCES

Most other uncertainties can be covered by "Allowances," where applicable. After completing the base estimate, estimators need to prepare a list of all the uncertainties, other than the "known" items already covered by our definition of contingency. They then must decide which, if any, of these uncertainties must be added to the estimate. The criterion they should use is to only consider an item if it is predicted that it is most probable the item will be needed or the event will occur (i.e., there is a better than 50/50 chance of it happening). If this is the case, then the "uncertainty" should be analyzed, a description or definition prepared, and an estimate made on this basis. The estimated allowance would be added to the base estimate as a specific allowance, identified and with a documented basis. If the uncertainty is judged to have less than a 50/50 chance of occurring (i.e., is unlikely to happen) it should be left out of the estimate. If concern still exists with regard to this item, an estimate can be made of the cost and schedule impact and a sensitivity check run on the project economics with the final decision as to whether to include or exclude left to management. Adding additional allowances for concerns such as strikes and bad weather is usually a subjective decision and biases an estimate on the high side and produces unrealistic estimates which are unsuitable either for funding or as a control tool. This practice should be avoided.

Major scope changes, weather extremes, accidents, acts of God, currency fluctuations, and other similar events should also be excluded from estimates. These items are unpredictable, both in terms of whether they will occur, and the magnitude of the cost and schedule impact on the project if they do occur. For a contractor, major scope changes are covered contractually as an extra to his estimate. For an owner, any major scope change is, in reality, a new project and requires a new estimate with new economics.

ADVANTAGES OF THE RECOMMENDED APPROACH

A number of significant advantages exist if the approach recommended is adopted:

1. It provides a consistent and rational approach to the estimating and control of estimate uncertainties, an area often laced with subjectivity and emotion.
2. It segregates known items which occur on every project, and therefore can be analyzed, from the more vague and more subjective unknown items, which may or may not happen.
3. By separately identifying, defining, and estimating each allowance, it prevents the area of uncertainty from being a source of "double-dipping" or duplication of contingencies and allowances during the preparation of the estimate. It also prevents this area from becoming a "slop fund" where overruns or underruns can be hidden.
4. Because of the consistency and the segregation of the unknown items, an analysis can be made of historical data and realistic contingency levels established, keyed to an estimate classification system, as discussed in Chap. 2.

CONTINGENCY CONTROL

As we have seen from the above discussion, contingency can and usually does constitute a significant part of the total cost in any estimate. As mentioned earlier, it can vary in amount from as high as 50% in estimates made in the early stages of project development to as little as 2 or 3% in estimates made at the final stages of a project. On average, the contingency added is usually in the neighborhood of 10 to 15%. Consequently, it is one of the most significant "one-line" items in an estimate and is certainly worthy of attention and some effort. This effort is not only in the way of spending the time to analyze the need and develop a rational basis for estimating the contingency required along the lines we have just discussed, but it also means that contingency should be monitored and controlled while the project is executed. Table 3 outlines the five steps that can be taken by the CE and project managment team during project execution to exercise a degree of control over contingency.

In the first place, for the contingency to be understood and therefore administrated properly by both the project management team and upper management, it is necessary that they have available concise documentation of the reasons the contingency was included and the basis for predicting the amount included. All people who in any way can influence the expenditure of the contingency should be educated along these lines. As discussed earlier, the

Table 3   Contingency Control

Document basis
Educate participants
Control changes
Forecast (reestimate) monthly
Do not treat as "slop fund"

most common reason for adding contingency is to cover the cost of the inevitable changes that are made to a project basis after the estimate is released. In a later chapter we will discuss in detail procedures that can be set up to monitor and control the number and value of these changes. For now, it is important to realize that this area is easily controlled provided the recommended systems are adopted and followed with management backing.

Lastly, the amount of contingency needed to complete the project can and should be forecasted or reestimated on a monthly basis in a manner similar to any other item in the estimate (i.e., a prediction should be made of future performance based on performance to date). Under no circumstances should contingency be treated as "slop fund," as is so often the case. It is not an item to be used to cover overruns, nor should underruns be covered up by increasing the contingency forecast. As we will see in later chapters, this type of approach makes it difficult to implement effective cost control and does a disservice to management by producing misleading forecasts and cost reports. Control of any allowances which are included as line items in the estimate can be exercised by reforecasting the item each month and eliminating the allowance from the monthly forecast, once the event has occurred. If the event occurs over a period of time rather than a single occurrence, then a run-down curve can be used as the basis for the monthly forecast and monitoring of the remaining allowance required.

SUMMARY

In this chapter we have touched lightly on three of the more troublesome areas for the CE: escalation, allowances, and contingency. We have learned that, with the proper approach, these problem areas can be handled in a rational way to produce both reasonably accurate and useful estimates.

In determining the amount of escalation required, we emphasized the fact that the actual selling price of an item (and therefore the predicted or estimated cost) is determined more by the degree of economic activity (competition) than the actual cost to the vendor of making the item.

The type of contingency to be added is determined to a large degree by the quality of the basis for the estimate and the estimating methods used. The actual amount is a function of the company's experience and the time in the development of the project when the estimate was done. An important fact to be considered when discussing contingency is that each company has its reasons for including a contingency and, hence, its definition of what contingency is. This specific definition must first be determined before any meaningful discussion on the subject can take place.

Because of its many advantages, an approach which segregates a contingency allowance for changes and inadequacies in methods and data from all other allowances is highly recommended.

10

ESTIMATING A REVAMP PROJECT

GENERAL

In all of our previous discussions of estimates, estimate accuracy, and estimating methods, we dealt with new or grassroots plants (i.e., projects that are started on a virgin site and therefore can follow the normal, optimum project execution sequence in both the engineering and construction phases). Because there are no abnormal constraints, sufficient data can be gathered and analyzed to allow the CE to develop reliable estimating methods to reflect or predict the average project execution performance that one can expect on a new project. Hence, all of the estimating methods discussed in Chap. 4 were based on this averaging concept. Of course, all projects are not new, since there are many economic or other incentives for making revisions to an existing plant. A project involving such revision is called a "revamp" project.

WHAT IS A REVAMP?

There are many examples of revamp projects. Some of these and an explanation of why they occur follow.

Debottlenecking Project

An owner finds it far more economical to expand an existing plant than to build a second grassroots plant. The lower incremental cost of expanding the plant and the lower incremental operating cost normally provide an attractive economical return. The expansion is called a "debottleneck" because the design for the incremental capacity is determined by reviewing the existing design and only increasing those areas where the existing design is incapable of handling the increased capacity. Surprisingly, this review reveals that only 50% or less of the equipment requires modification. The reason that the remainder of the equipment is found to be adequate for the larger capacity is that it is common practice for owners and engineers to build some "fat" or conservatism in the original design. Also, many equipment items tend to be sized or manufactured in stepwise fashion. In going to the next available or desired size, some extra capacity is automatically built into the original design. A debottlenecking project, in most cases, involves going into the guts or heart of a plant and making removals, relocations, and replacements.

Internal Additions

On occasion, the designer will find it necessary to add additional steps to the process, such as additional treating or fractionation facilities. For operation and economic reasons, it is usually desirable to locate these new pieces of equipment in the existing plant. Although not as complex as a debottleneck, this internal addition can be tricky, particularly if the new tower or drum must be set in the middle of the plant.

Plant Modernization

New plants, like people, grow old eventually. Unfortunately, unlike people, they become less efficient and less in tune with modern technology as they grow older. However, usually the heart of the plant is sound and worth fixing up or modernizing. Modernization is done not only to make the plant more efficient but often to update the safety requirements or to meet new and more stringent local environmental regulations. A plant modernization, like a debottleneck, can be complex and difficult.

WHY ARE REVAMPS SO DIFFICULT?

By their nature, revamp jobs do not lend themselves to good, concise definition. They are not "clean" and, particularly in the early stages, the job definition is vague. Hence, the CE is presented with a poor basis on which to make an estimate. Even if he had a better definition, the CE still would be faced with trying to predict a job execution that is nonoptimum, does not follow the general orthodox approach, and by all standards is definitely not average. Consequently, it is not surprising that estimates of revamp projects overrun. By examining some of the specific reasons that these estimates overrun, we can gain a better insight into the problems the CE encounters in preparing the estimate.

Table 1 Typical Revamp Overrun

Existing equipment inadequate
No plant shutdown for construction
Bulk material overrun
Underestimate of engineering, especially field
Inadequate staging/access
Excessive overtime/other premiums
Poor supervision
Poor productivity

TYPICAL REVAMP OVERRUN

Table 1 lists the major reasons that many revamp estimates overrun. Note that the list is a mixed bag covering poor estimate basis (i.e., definition of equipment requirements, plant availability for tie-ins), poor estimate or estimating methods (i.e., bulk materials, engineering, and FLOH), and poor actual performance (i.e., supervision and productivity). In other words, just about anything that could possibly go wrong, can and often does, unfortunately. To overcome this problem and produce a more reliable, consistent estimate, the CE needs to develop an approach that is different from the averaging concept we have been applying to new projects. He not only needs to adjust his estimating methods to suit these nonaverage conditions, but he also needs to expend more effort in developing the basis and in preparing the estimate. Let us first analyze what is needed overall to produce a good estimate of a revamp project.

INGREDIENTS FOR A GOOD REVAMP ESTIMATE

Table 2 lists the five essential ingredients that the CE and the project people must have to produce a reliable estimate. Note that the first reason for an overrun listed in Table 1 is the inadequacy of the existing equipment to do the job required of it for the new design basis (i.e., the expanded or debottlenecked

Table 2 Ingredients for a Good Revamp Estimate

Current, accurate plant history
Firm engineering basis
Realistic assessment of field performance
Detailed look at schedule
Designer/owner/operator involvement

design rates). Often the designer does not take the time out to check in detail with the plant operators and thereby learn what the current capacity and conditions or the existing equipment are. To get this information, it is necessary to conduct test runs of the plant or analyze in detail recent inspection and maintenance reports on the individual pieces of equipment. By securing an up-to-date plant history, the designer eliminates a lot of the guesswork and minimizes the number of times she leaves open questions in the design specs. Statements such as "actual conditions are to be checked at the time of the shutdown" are of little help to the CE. If she assumes the worst, she could overestimate and kill the project economically before it starts. If she is overly optimistic and assumes the equipment is okay, we are potentially back to our overrun problem. What she needs is the correct answer beforehand, and this can be provided by a current plant history.

Up-to-date information on the existing plant will also provide the designer and CE with the latest plot plan and layout arrangement and the whereabouts of the underground facilities. These are important considerations to both basic and detailed engineering and, hence, to the CE in making her predictions in the estimate.

FIRM ENGINEERING BASIS

The design specifications prepared by the owner provides the contractor with the basis for his detailed-engineering work. If the specification is based on many "open" items regarding plant capacity or the condition of the existing equipment, or if it contains many assumptions based on old plant history, many problems will surface during the detailed-engineering phase. Not only is a good plant history essential but an update of all other requirements is needed. This includes looking into such areas as the most recent local safety and environmental requirements. Also essential is a much greater involvement at a much earlier time of the owner's plant operators in the development of the design basis. Above all, the results of their analysis and their recommendations should be realistic. Unfortunately, since this realistic information is normally required at the time when everyone is trying to produce as low an estimate as possible to keep the economics favorable and secure project approval, there is a conflict. In general, there's a tendency to put on rose-colored glasses and come up with minimum or below-minimum requirements. It's up to the CE and management to insure that this doesn't happen.

ESTIMATING FIELD LABOR IS DIFFICULT

For many reasons, predicting the number of manhours required of field labor on a revamp project is difficult. As mentioned earlier, the work is not carried out in an optimum sequence, since part of the project already exists. In addition, the work is carried out under adverse conditions. Congested work areas are common and much of the work is done at the higher elevations (in existing pipe racks and on existing steel structures). Much of the work must also be done while the plant is in operation and, for safety reasons, many restrictions, such as "hot-work" permits, are imposed. Productive time is also lost in going to special smoking areas or as a result of gas alarms if a dangerous gas leak occurs in the existing plant during operations.

Experience has shown that anywhere from 10 to 50% more manhours may be required to complete the same task on a revamp compared with a new project, depending on the severity of the above conditions. The CE must analyze the entire situation in detail and predict what the conditions will be and what adjustments must be made to the normal "average" manpower requirements.

The degree and nature of the adjustments required can be calculated beforehand in the form of an estimating method, as discussed in Chap. 4. This can be done by retaining the method developed from averages (for a new plant) as a base and increasing or adjusting the level of manhours required because of revamp complexity. Figure 1 illustrates one way of doing this. In the revamp

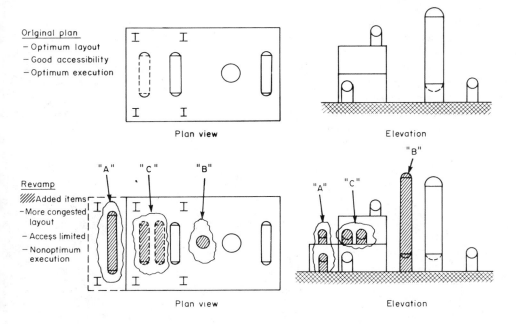

Figure 1   Revamp project area classifications.

figure, all areas to be modified have been identified and given an "area classification" according to the complexity or difficulty of installation (i.e., A, B, and C). A represents the addition of equipment adjacent to the existing plant but still grassroots from the standpoint of access and erection sequence. However, this classification still carries some debits for such things as the need for hot-work permits and a no-smoking area. Areas B and C are more difficult, with equipment installed in the middle of the existing area. These areas have all the problems of a revamp project (i.e., congestion, poor access, difficult erection, etc.). In particular, area C is complex because of the above-grade work in an existing structure. In the more difficult areas, such as B or C, significant increases will be made in the manhour estimate over what would be allowed for a new plant. The amount of percentage addition would be determined from an analysis of past revamp jobs for work done in areas of that classification. In addition to these increases to reflect difficulty of installation, other adjustments must be made for the nonproductive time (i.e., smoking breaks, gas alarms, etc.). This would also be done based on past historical experience.

SCHEDULING IS A KEY ITEM

In any estimate, scheduling is a key factor. The start and length of the project schedule determines such major cost items as escalation and the cost of FLOH. If the schedule is not of average length (i.e., it is to be shortened purposefully) then other cost factors enter the picture, such as overtime premiums and the loss of productivity associated with the overtime.

A revamp project is, as we have seen, definitely not average and, hence, the schedule for each project must be developed by the CE tailor-made for the project. For virtually all revamp projects, a major number of tie-ins to existing equipment will be required. In addition, there is the replacement of existing equipment with new equipment or the relocation of piping. To accomplish these tasks, it is necessary to shutdown the existing plant. To avoid a special shutdown for the project, these tasks are carried out during a plant "turnaround," (i.e., a periodic, scheduled shutdown of the plant for maintenance and inspection). By making the tie-ins during this period, the company saves the considerable loss that could result from lost production during a second shutdown. However, it then becomes extremely important that the revamp project be scheduled in such a way that all work that can be done during a shutdown be done at the time of the turnaround. Also, the amount of work must be minimized and scheduled to fit the time frame of the turnaround; otherwise, any lengthening of the turnaround and subsequent production loss will accrue as a debit to the project economics. To meet this kind of schedule, extra costs and additional problems exist that must be predicted and reflected in the estimate by the CE.

For example, most turnarounds are run on a 24-hr basis in two or three shifts. Shift work involves additional costs in the form of shift-wage premiums, additional supervision, floodlighting, other overheads, and lost productivity due to overlap problems. Additional premiums are also paid to insure that material is fabricated and delivered on time (i.e., before the turnaround starts). All of these potential increased costs must be considered and included where applicable. In addition, because of work density or congestion, scheduling is much more difficult. This combined with the reasons mentioned above makes scheduling one of the most critical areas in a revamp project. The CE needs to recognize the job that has to be done in this area, its limitations, and the average cost impact. In other words, both he and the project management team must spend more time establishing the schedule for the estimate basis on a revamp than they normally would for a new, grassroots plant.

MATERIAL COSTS ALSO GO UP

Direct-material costs on a revamp project tend to be higher than a comparable new project by as much as 5 to 10%. Part of the reason for this is the nonoptimum nature of a revamp project's layout. Piping, for instance, cannot be laid out from scratch in a average, optimum way. The designer is forced to route the piping where there is space available, on or alongside existing pipe racks or supports. This results in more fittings and a more complex, more extensive piping layout. New foundations have to be squeezed into the available underground space, straddling existing undergound obstructions, at a higher cost. More material than average is required also because tie-ins disturb existing supports, insulation, paint, and much of this has to be replaced or repaired. It is not uncommon for an owner to request a complete repainting of the plant, especially if the revamp has been extensive. In addition to the cost of this material, the labor to install the incremental material is a further debit to a revamp project that the CE must consider.

FIELD LABOR OVERHEAD AFFECTED

The amount of FLOH required per manhour increases significantly in a revamp project. Because of the complex nature of the work, a higher ratio of supervision to workers is needed. Most of the work is above grade and this tends to require more expensive support construction equipment, such as cranes and welding machines. Additional scaffolding is also needed to work around and above the existing equipment. Clean-up or housekeeping becomes more expensive because, for safety reasons, the operating plant must be kept clear of construction debris on a daily basis. "Plus rates" (i.e., premiums paid in some areas to labor for adverse working conditions) can add costs for work inside existing towers or at high elevations. All in all, FLOH for a revamp job can be as much as 25% above the normal average for a new project.

ENGINEERING BECOMES MORE COMPLEX

Detailed engineering on a revamp involves more manhours and hence more costs than a new project. Engineering checks have to be made of all the existing facilities to insure they are adequate and to put the designer in a position to produce an integrated design, new and old. This not only involves more engineering manhours per piece of new equipment but can entail weeks of work in the field establishing detailed locations of piping routings, tie-ins, access, etc. Many contractors elect to do all of the detailed engineering in a field office to be closer to the problem areas. Although this can be justified since it shortens the lines of communication and minimizes errors, it results in higher overhead costs than an equivalent new plant, and these increased costs must be reflected in the estimate. Overall, detailed-engineering costs on a revamp project can run as high as 50% above a grassroots plant.

REVAMP PANEL

One step that can be taken to insure the estimator has the most realistic basis for the revamp estimate is to create a revamp committee or panel prior to finalizing the estimate basis. The panel would be composed of people who have not necessarily been involved full-time on the development of the project but who have the background and expertise to critically review and comment on the proposed estimate basis. Normally the team would consist of an estimator, a designer, a construction engineer, a safety expert, and a layout engineer. They would spend up to a week reviewing the design and execution basis. Much of their time would be spent at the site. The report of their findings would be incorporated in the estimate basis. In this way, many of the items which are omitted or given light treatment in the basis would be "flushed out" by the revamp team and included in the estimate, rather than showing up as an overrun later.

CONTINGENCY ALSO NEEDS TO BE INCREASED

As we learned in Chap. 9, contingency is added to cover changes to the estimate basis, methods and data. Since there is a higher degree of uncertainty and variability associated with both the estimating basis and estimating methods for a revamp project, there is justification for considering a higher contingency.

Table 3   Revamps: Summary of What Is Needed

Latest plant data
Field inspection
Owner involvement
Thought-out basis
Revamp panel review
Detailed schedule
Method geared to complexity

Most of the uncertainty and variability historically shows up as omissions or inadequate design basis, so that the estimate level tends to be biased lower than its grassroots counterpart. Increases of 50 to 100% in contingency level (i.e., from 10 to 15 or 20%) are common.

SUMMARY

Table 3 summarizes the areas that must be given special attention when the CE is establishing the basis for and preparing his estimate of the cost of the revamp project. As is often the case, the most important consideration is getting the people who are involved and in a position to know to stop long enough to give the CE a realistic prediction or basis for executing the project. This, combined with a rational approach to adjusting the estimating methods to a nonaverage situation, can give the desired result of an accurate and useful estimate.

Part II
COST CONTROL

11

INTRODUCTION TO COST CONTROL

BASIC RESPONSIBILITIES OF PROJECT MANAGEMENT

In the planning and construction of a new grassroots process plant or an addition to an existing plant, the project management has the following three basic responsibilities:

1. Quality control: The plant and its parts must be designed and constructed to the owner's quality standards. The plant should be built to operate safely for a specified number of years producing the specified products.
2. Schedule control: The plant should be completed and onstream at the specified time so that the owner can meet product delivery commitments to his customers.
3. Cost control: The plant should be completed within the budget so that the owner can realize his expected profits and keep financial requirements within anticipated limits.

These three responsibilities are equally important and to some degree interdependent. For example, quality can usually be improved with an expenditure of additional money and, on most projects, schedule can be improved if one is willing to pay premiums to equipment vendors and overtime to construction workers. It is project management's job to see that equilibrium is maintained between quality, schedule, and cost.

OBJECTIVES OF A COST CONTROL PROGRAM

A project cost control program has the following four objectives:

1. To focus management attention on potential cost trouble spots in time for corrective or cost-minimizing action (i.e., detect potential budget overruns before rather than after they occur).
2. To keep each project supervisor informed of the budget for his area of responsibility and how his expenditure performance compares to that budget.
3. To create a cost-conscious atmosphere so that all persons working on a project will be cost-conscious and aware of how their activities impact on the project cost.
4. To minimize project costs by looking at all activities from a cost reduction point of view.

COST REPORTING vs. COST CONTROL

Cost reporting must not be confused with cost control. In the past there has been a tendency for project managers (PMs) to issue a monthly cost report comparing actual expenditures to the inital estimated cost of the project and labeling this cost control. It is not control, but reporting. Reporting actual expenditures is the accountant's job. Forecasting cost trouble spots before funds are committed and determining corrective action to minimize these expenditures is the CE's job. In other words, cost control is spotting trouble spots and taking action, and the accountant's cost report is one of the tools used by the CE.

ELEMENTS OF A COST CONTROL SYSTEM

For any engineering/construction project, the elements of a good cost control system comprise the following five points:

1. A planned approach to the project: To realize maximum economy all activities must be carefully planned as to timing and method of execution.
2. A realistic financial yardstick: This is the control (budget) estimate.
3. Accurate and timely cost forecasts: These cover the costs to completion for all activities.
4. Comparison of forecasts to the yardstick: This is a detailed item-by-item comparison of forecasted costs with the budget.
5. Positive action to minimize forecasted budget overruns, the essential ingredient of cost control.

THE CONTROL ESTIMATE

The control or budget estimate is the main tool of the CE. This estimate is a detailed prediction of the project execution plan. It reflects not only the schedule but also the physical and economic conditions under which the project will be executed. The estimate must be in sufficient detail to permit actual project performance to be measured against it on an item-by-item basis.

The budget must be available when needed. For example, if an important long-delivery piece of equipment must be ordered early in the project, then the estimate for that piece of equipment must also be prepared early and be on hand for comparison with the vendors' bids when they are opened.

PROJECT CHANGE ORDERS

Even what seems to be the most logical approach and best thought-out design will need modifications as work progresses. It is always necessary to make changes during the execution of any engineering/construction project. For effective cost control, it is necessary to alter the project budget to reflect changes as they occur, and the best way to do this is with a formal change order system. As each change is decided upon, its impact on cost and schedule is determined and incorporated into the project budget. Each change should be documented on a special form and formally approved by the PM. By use of the change order system, the project budget is kept up-to-date, always reflecting the current execution plans. The change order system will be discussed in greater detail in Chap. 18.

THE NEED FOR WRITTEN PROCEDURES

In the engineering and construction fields, cost is the one item that can be neglected until too late for corrective action. Commitments are made to fabricators of process equipment many months before that equipment is delivered and the invoice paid. In process plant construction, the accountant-reported expenditures lag actual commitments by many months; in the case of sophisticated equipment the lag can be as great as 1 to 2 years. Also, in periods of high inflation rates, equipment fabricators cannot quote fixed prices but quote a base price plus an escalation to time of delivery. The escalation rate is usually pegged to some common indicator, such as the U.S. government's Wholesale Price Index. On a large process plant project many commitments are made simultaneously and various engineering specialists are involved. It is easy to see that without a strict, formalized cost control system, commitments far in excess of available budgeted funds can be made before anyone is fully aware of the disaster. Situations like this frequently cause project cancellations at great financial loss or bankruptcy to the owners.

Situations of the kind just described are best avoided by a formalized approach as outlined in a written cost control procedure to be followed by all project personnel. This written procedure establishes clear-cut cost responsibilities and lines of communication. The procedure covers all aspects of cost control including methods for handling or preparing the following:

Tabulations of fabricators' bids and comparisons with the budget
Approval of making commitments
Project change orders
Forecasts of costs to project completion
Disseminating cost information to all concerned
Format for tabulating, forecasting, and reporting project commitments

TYPES OF CONTRACTS

The following three distinct types of construction contracts are commonly used in the process plant construction industry:

1. Lump sum: The contractor assumes the risk of overrunning his budget and losing money. In periods of low economic activity when inflation is not a problem, contractors are willing to bid lump sum for small well-defined projects. In this situation, the owner needs only a part-time CE to control and estimate changes.
2. Fully reimbursable with fixed fee: The contractor is reimbursed for all project costs and is paid a fee for his services. With this arrangement the owner's CE carries a heavy responsibility for cost control, and he will be deeply involved in all of the contractor's activities.
3. Fully reimbursable with a fixed-target cost and mutual sharing between contractor and owner of any underrun to that target cost: The contractor is usually paid a minimum fee but has a strong motivation to spend the owner's money wisely and underrun the target cost, thereby increasing his own profits. In contracts of this type, responsibility is shared between the owner's and contractor's CEs.

There are various combinations of these three contract types. One frequent combination is fully reimbursable engineering with construction under a nego-tiated lump sum. Many other variations are encountered, but in all cases the professional CE strives to keep costs within the budget and to minimize expendi-tures in keeping with quality and schedule requirements.

WHO CONTROLS COSTS?

Costs can only be controlled by those who are doing the spending. In all usual situations this is the contractor. As we have seen, control of costs is one of his three prime responsibilities, the other two being quality and schedule. However, the owner does have considerable impact in this area since he is directly respon-sible for the basic design specifications and for changes made during project execution. Also, if an owner wants to be sure that he is getting good cost control on a reimbursable contract, he must monitor and appraise the contractor's cost control efforts. In other words, the owner must enter into the details of the contractor's work in sufficient depth to be sure that good cost control is exercised.

HISTORICAL PROJECT PHASES

Historically, process plant projects evolve through four fairly distinct phases, as illustrated in Chap. 1, Fig. 1. The duration of these phases varies significantly from project to project, and the phases themselves frequently overlap. The time intervals and overlaps shown in Chap. 1, Fig. 1 are order of magnitude only and not intended to represent specific projects. Nevertheless, these distinct phases nearly always exist, and normally, different people are involved in each phase. We can think of cost control as divided into similar phases listed below:

Cost control during conceptual engineering
Cost control during detailed engineering
Cost control during construction

The specific duties of the CE and control methods differ for each phase, and in later chapters we will examine each phase in detail.

EXAMPLE OF COST CONTROL

The technique most frequently used by the engineer to spot potential cost problems compares that which is actually happening with the predictions of the control budget. Isolating and correcting of a cost problem includes the following steps:

1.   Comparison: locates the problem
2.   Forecast: predicts magnitude of the problem
3.   Analysis: determines reason for the problem
4.   Corrective action: selects a lower cost alternative
5.   Revised forecast: reflects impact of the corrective action on project costs

The simple example shown in the following Figures 1 and 2 illustrates an application of the preceding five steps.

Execution plan: 2 stores, 4 items:
Budget: allowance = $100
Commitments: after 1 store, 2 items:
                          Spent = $50
                          Remaining = $50
                                   Total $100
                          OK(?)

Figure 1   Shopping budget: poor cost control.

Items	Budget	Spent	Left	Forecast	Over/under
Dress	50	—	50	50	—
Hat	20	—	20	20	—
Shoes	20	35	—	35	+15
Gloves	10	15	—	15	+ 5
Total	100	50	70	120	+20

Analysis: poor purchasing (but average performance)
Corrective action: buy material; make dress = $20
Revised forecast: $90 vs. $100 budget

Figure 2   Good cost control.

WHAT IS NEEDED FOR EFFECTIVE COST CONTROL?

If cost control is to be effective, the following ingredients must be present within the project management organization:

Management attitude that emphasizes cost control
Cost-conscious design/construction team
Capable cost control organization
Full-time cost follow-up
Good cost tools
Comprehensive written procedures that formalize the approach to cost control

SUMMARY

The CE is the focal point of the cost control program. His objective is to minimize project costs, and he does this by focusing management's attention on potential cost trouble spots in time for corrective action. His main tool is the control budget, and his approach is to analyze existing trends and forecast total costs. These forecasted costs are compared to the budget and potential overrun areas called to management's attention for corrective action.

In the next chapter we will see how large-process-plant engineering/ construction projects are managed and how the CE fits into the overall project management team.

12

THE COST ENGINEER AS PART OF PROJECT MANAGEMENT

COST CONTROL: A PROJECT TEAM RESPONSIBILITY

As we have seen, cost control is one of project management's three responsibilities, and the CE is an integral part of the management team. Only those individuals making day-by-day decisions that affect cost can exercise any degree of control, and the CE provides these individuals (engineering and construction supervisors, specialists, buyers, and schedulers) with the tools and information needed for them to exercise control in their respective spheres of responsibility. Responsibility for cost control starts at all levels of endeavor. Each individual responsible for any phase of a project must be continually informed as to how their costs compare to budgeted amounts, and it is the job of the CE to keep them informed.

WHAT THE CE DOES

On a project management team, the CE is the focal point of cost control. Her specific duties can be categorized as follows:

Provides project management with the information needed to control costs
Points out areas of cost overrun
Investigates and recommends corrective action
Forecasts costs and prepares a complete project cost outlook on a monthly basis
Monitors costs between outlooks and immediately advises the PM of any event
    or decision that has a significant cost impact
Keeps a complete record of changes to the project
Provides quick cost estimates for design alternatives

Sends cost information back to home office or central file for use in future
estimates

THE ENGINEERING ORGANIZATION

Most engineering contractors engaged in process plant design and construction
are organized in a manner similar to that shown in Fig. 1. The engineering organization is generally headed by a vice-president, division head, or individual with
a similar title. Reporting to this individual, as shown in Fig. 1, are the following
four major departments:

1. Purchasing: Includes those individuals who actually buy and expedite
    materials from fabricators and subvendors.
2. Engineering: The backbone of the organization. Includes designers, tech-
    nical specialists, and draftsmen. Of the total technical personnel, 50 to
    75% are in this department.
3. Project control: Includes cost and schedule engineers. Some organizations
    consider estimators separate from engineers engaged in cost control, but
    when this is true, there is frequently a great deal of personnel transfer
    between the two groups. Schedule and cost engineers are usually separate

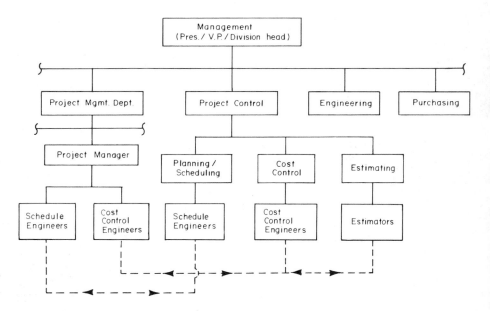

Figure 1   Ideal organization.

and distinct groups but their work is so interrelated that they must maintain close liaison. This department normally lends individual engineers out to work on specific project teams.

4.   Project management: Consists of a pool of project engineers (PEs) and managers (PMs). These people are normally assigned to specific project teams, and they are the persons who generally maintain direct relations with clients and owners. Their functions combine management, coordination, and client relations.

THE PROJECT MANAGEMENT TEAM

Most contractors use the task force approach to project management. When a contractor or consultant receives a commission to perform engineering, he will normally assign a PM to take overall responsibility. The PM will then choose a team of PCs and specialists as assistants. This management group, varying in size from 1 to 2 persons for a small project to 40 to 50 persons for a large project, will have the responsibility of providing designs, detailed drawings, material procurement services, schedules, cost estimates, and cost control throughout the engineering phase of a project.

Designers, draftsmen, and other technical personnel are assigned to the task force as required. These individuals generally remain under the overall

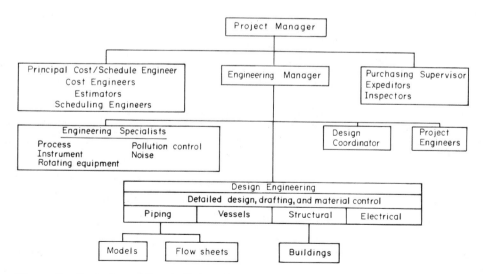

Figure 2   Task force, South Pole refinery. Approximate peak, 500 persons.

control of their respective department heads; however, their services are allocated to the task force and they perform their duties under the directives and schedules of the PM.

Figure 2 is a typical task force organization as used to engineer a project requiring a million technical manhours or more. The organization for a small project, less than 200,000 technical manhours, would be modified and appear as shown in Fig. 3. As can be seen from the two figures, the task force must perform the same functions for both large and small projects. The main differences are the number of persons involved and ranking of the management personnel. For example, the PM assigned to the large project will be a senior, experienced individual, whereas the PM on the small project may be a younger individual with less experience. On the large project, cost and scheduling will be handled by a group of 4 to 10 persons, while on the small project only 1 or 2 persons will be required.

INTERFACE BETWEEN COST CONTROL
AND OTHER ENGINEERING FUNCTIONS

In his daily work the CE contacts nearly all other members of the task force. He is in daily contact with the PM, works hand-in-hand with the scheduler, discusses costs and budgets with the engineering specialists, maintains close

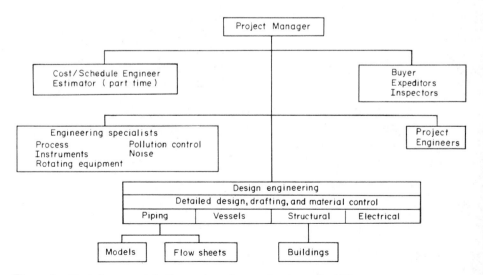

Figure 3   Task force, catalytic cracker. Approximate peak, 100 persons.

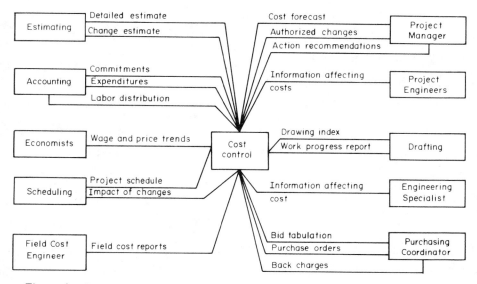

Figure 4   Cost control flow chart.

liaison with the purchasing coordinator, and is the task force's main link with accounting. In addition, a contractor's CE will have direct contact with the client, and the client's CE will have many contacts with all members of the contractor's task force. The CE must be a knowledgeable, versatile individual who can understand the problems of the instrument specialist as well as those of the piping designer. Figure 4 is a graphic illustration showing the CE's relationship to other members of the task force.

THE OWNER'S ORGANIZATION

Whether a contract is a lump sum or reimbursable, the owner will have one or more engineers resident in the contractor's office throughout the engineering phase. On large reimbursable projects, most owners assign a fully staffed project team as shown in Fig. 5. In an organization of this type, the owner's PM has overall responsibility for the entire project, and he is assisted by a team of specialists, each of whom monitors the contractor's work in his area of expertise. The purpose of this team is to interpret the owner's specifications and requirements to the contractor and to verify that the contractor's designs accurately reflect the owner's intentions. Also, in a reimbursable situation, the owner wants to verify that the contractor's working methods are efficient and

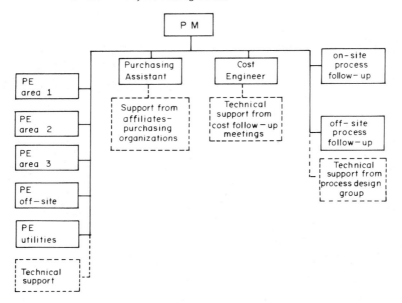

Figure 5   Owners's PM team, contractor's office.

that designs are economical. In short, the owner wants to make sure that his money is wisely spent by the contractor.

Referring to Fig. 5, the owner's PM is assisted by several PEs, each responsible for an area or specific facility. The PEs are in daily communication with their opposite numbers on the contractor's task force and verify that the detailed design meets the owner's requirements in all aspects. Also, they draw technical support by referring questions back to their home office and relaying the answers to the contractor.

If the owner has an existing plant, he normally has a purchasing organization and preferred vendors that supply equipment. Also, if the owner has a large company, his purchasing expertise and market leverage may be equal to or greater than the contractor's. Therefore, the owner's PM team usually includes a purchasing assistant who works closely with the contractor, thus combining the purchasing facilities of both organizations to get the best possible prices and delivery for the sophisticated equipment required in process plants.

Normally the owner's PM will include one or more process engineers on his staff. This is especially true if the owner has developed and patented the process or if he has contributed substantially to development of the process. These individuals will usually be resident in the contractor's office until completion and approval of all piping and instrument (P&I) flow sheets. After approval of the P&I diagrams, the process engineers frequently return to their home office but remain on call as consultants to the PM.

The final member of the owner's PM staff is the CE, who frequently follows scheduling as well. The CE is a key peron who, like the PM, has an interest in the work of all other staff members. Also, the CE must keep abreast of recent trends in both material and labor escalation. Frequently he is supported in this by follow-up reports received from other projects in which his company may be engaged and by information gathered from the company's purchasing department. He pools this information with that gathered by the contractor and, working closely with the contractor's CE, advises the PM of the escalation outlook for his particular project. The owner's CE is in daily contact with his PM and contractor's CE. Also, he has frequent meetings with all key members of the contractor's task force.

On a small project the owner may be represented only by the PM and one or two PEs. Frequently one of the PEs handles the CE's function on a part-time basis. In this situation it is wise of the owner to have an experienced CE make periodic visits to the contractor's office to lend support and guidance in cost areas. On a lump sum project the owner is concerned with quality only and his team will probably not include either a CE or purchasing assistant.

HOW COST CONTROL WORKS

Cost control starts at the top—with the PM. With a PM's support, it is the job of the CE to instill a sense of cost awareness into every member of the project team, both owners and contractors. Each task force member must be a practicing CE in his specialty and this must be stressed to him by the PM. It is the CE's job to provide the tools, information, and guidance in the cost area that enables the specialists to exercise control.

COST CONTROL MEETINGS

One of the best methods of instilling cost awareness throughout the task force is to hold regularly scheduled cost control meetings (i.e., meetings where the subject is cost and what is being done about it). All supervisory members of the task force should attend these meetings and each person should be required to discuss costs within his area of responsibility. Subjects to be discussed by each person include the following:

Results of current corrective action that he is now taking
Recently discovered areas of budget overrun
Corrective action that he intends to take in an effort to minimize costs

Nothing makes a supervisor more cost-conscious than knowing that he is held strictly responsible for costs in his area and that he must discuss the cost status at weekly meetings. When the supervisors are cost-conscious, the feeling of cost awareness soon spreads to all members of the task force and we then have cost control. The CE, working through a formalized cost program, is the catalyst that makes cost control work. He focuses attention on potential cost trouble spots, but he must rely on others to take the specific actions necessary to provide effective cost control.

CHANGES TO THE PROJECT

From the CE's point of view, a change to the project is any execution or economic development that is different from the basis of the current control estimate. Regardless of what stage of development the project is in, some type of estimate has been made, and it is the CE's job to evaluate the cost and schedule impact of each change to that estimate basis. If the project is in an advanced stage of development with an established definitive estimate, a formal change authorization procedure should be set up and authorizations above a specified limit (say, $500) should require approval of the owner's PM. The CE is the focal point of change orders. He numbers and keeps an up-to-date log of all change orders. Whenever a change is contemplated by any member of the task force, he advises the CE, who prepares a change order form immediately and starts it on its route to formal approval or disapproval. The task force operates under a strict formal change order procedure and the status of changes, both pending and approved, is discussed at the scheduled cost control meetings.

OVERALL QUALIFICATIONS OF THE CE

The CE must keep informed about all aspects of the project because, like the PM, he views the project in its entirety. Therefore, he needs a broad technical background. He must be able to intelligently discuss problems with specialists, such as instrument and electrical engineers, structural and piping designers, scheduling engineers, and purchasing agents. Previous experience as a project or field engineer is a great help to the CE. Also, the CE must develop an interest in nontechnical matters such as labor productivity, union problems, overtime pay, construction tools, warehousing, and the many other items that make up total project cost. The CE is constantly giving information to and receiving information from other people and he must aggressively pursue the information he needs. In short, the CE must not only have a solid technical backgound but must be able to communicate successfully with all other members of the task force.

THE CE's ROLE DURING CONSTRUCTION

We have seen how the CE fits into management of the engineering task force. He has a similar role during construction. The contractor's engineering task

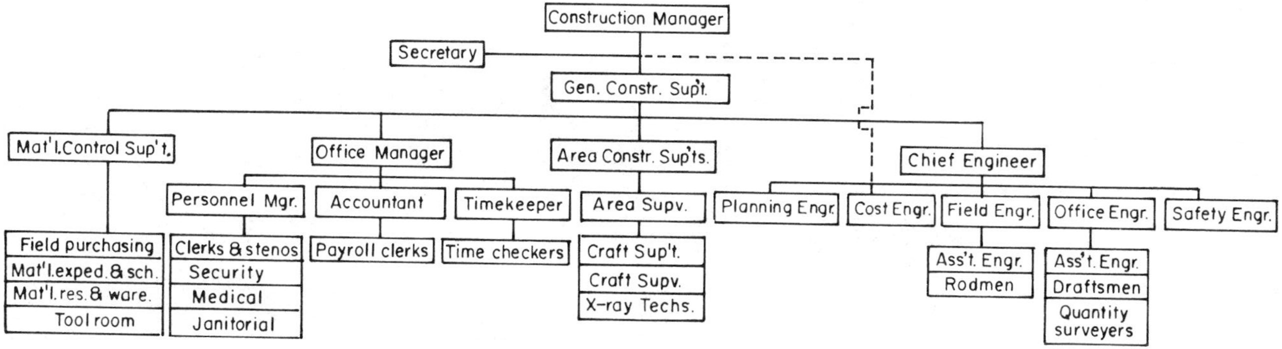

Figure 6 Typical field supervision organization. Chart is typical for a large refinery project. With decrease in project scope, some supervisory positions will be eliminated completely and responsibilities of other supervisory positions combined and handled by a single person. However, basic organization setup will essentially remain intact.

force will be replaced by a construction organization similar to that shown in Fig. 6. PEs of the engineering task force give way to construction area superintendents and engineering specialists are replaced by craft superintendents. On the owner's side, referring to Fig. 5, the PEs will be replaced by field engineers and the process engineers by start-up personnel. Although the problems will be different, there will be no change in the basic approach of either the owner's or contractor's CEs. Cost control during both the engineering and construction phases will be covered in later chapters.

SUMMARY

We have discussed the task force approach to project management and seen how the CE fits into the management team. The CE is concerned with all aspects of the project and frequently, like the PM, relocates from the engineering office to the field for the construction phase. During both phases it is essential that the CE have direct access to the PM. Any organizational setup that eliminates this direct reporting is endangering the cost control function. In real-life organizations, specialists and supervisors frequently try to convince the PM that their pet whim or proposal is the only way to solve the problem. It is the CE's job to impartially analyze the cost and schedule impact of these proposals and to lay the economic facts before the PM. In carrying out the analyzing function, the CE often comes across modifications or alternate proposals of his own, and he must be careful not to unconsciously distort the cost analysis to favor his own ideas.

In the next chapter we will begin our detailed discussion of how cost control is carried out in the offices of major engineering contractors and owner companies. We will start with the first of the three major phases in project development (i.e., cost control during the conceptual-engineering phase).

13

COST CONTROL DURING CONCEPTUAL ENGINEERING

INTRODUCTION

In Chapter 1 we saw that the planning and construction of process plants normally consists of three distinct phases where cost control can be applied. Referring to Chap. 1, Fig. 1, these phases are

1. Conceptual engineering
2. Detailed engineering
3. Construction

Cost control begins in the owner's office during the planning and conceptual-engineering phases. It is important to realize that the decisions made during the conceptual stage set the theoretical minimum cost of the project. For example, once the number of days of product storage is set, tankage capacity and a corresponding minimum cost become fixed. Similarly, once it is decided that two shiploading berths are required and that a tanker must be loaded in 10 hr, the minimum cost of the marine facilities is established.

The options open to management during the conceptual phase are greater than at any other time during the project. Hence, this phase offers the greatest opportunity for cost savings. Once the design basis has been set, the minimum investment is established.

Generally, the design basis, set during the conceptual phase, is not questioned throughout the remainder of the project. The pressures of a tight budget and schedule cause those engaged in the project to forge ahead without returning to an evaluation of the basis. For example, during detailed engineering it might occur to a piping designer that a 12-hour tanker-loading period would save considerable money in pumps and piping. However, the designer will probably not pursue this idea if the design basis has previously set the limit at 10 hr.

COST ESTIMATES AND BLACKOUT PERIODS

As a project passes through the phases of development, from evaluation and planning to construction, different types of estimates are made that apply to each phase. Although in industry various names are applied to these estimates, they fall into broad categories and, referring to Chap. 1, Fig. 1, we have labeled them as follows:

1. Screening studies: Early estimates used to compare competitive processes and possible plant locations. Screening studies are used during the evaluation and planning phase.
2. Preliminary estimate: The evaluation and planning phase culminates in the preliminary estimate, used as the basis for preliminary economics.
3. Semidetailed estimate: Prepared near the end of conceptual engineering. This is the estimate that owners frequently use to appropriate funds for a project.
4. Definitive estimate: A detailed estimate prepared as early as possible during detailed engineering. This estimate becomes the budget and is used as the basis for cost control during the remainder of the project.

Between these various estimates, progressing from preliminary through definitive, there are periods of cost blackout as shown in Fig. 1.

Preliminary economics (the reason for the project) are generally based on the preliminary cost estimate. If economics look favorable, the project moves into the conceptual phase. As conceptual engineering progresses, many changes are made to the original planning basis. As soon as this happens, the owner enters into a cost blackout period, as illustrated in Fig. 1. He no longer has a handle on the project cost. Often this blackout period will last for months until a semidetailed estimate can be prepared, and the cost may increase significantly, say, ± 50%, without management having an opportunity to take cost reduction steps or review the economics.

Initial concepts	
Screening studies	
Preliminary estimate ———————	Cost blackout
Conceptual engineering	3-6 months
Semidetailed estimate —————	Cost blackout
Complete conceptual engineering	
Start detailed engineering	3-9 months
Definitive estimate —————	

Figure 1   Cost blackout periods.

After the semidetailed estimate, assuming the economic outlook is still favorable, the project moves on into completion of conceptual engineering and the start of detailed engineering. Also, it moves into the second cost blackout period. This blackout period can be costly to the owner, because funds are committed at an accelerating rate; greater numbers of engineers and technicians are involved and commitments are made to suppliers of long-delivery equipment such as reactors and furnaces. The next chance to review economics comes with the definitive estimate, which might easily be 25% above the semidetailed version. If the project is now uneconomic or if the cost is above the owner's ability to finance, he will be forced to cancel at great monetary loss. All of the engineering cost will be lost, and a large percentage of the funds committed to vendors will be required for payment of cancellation charges.

COST MONITORING

The objectives of cost control during the conceptual phase are the following:

Eliminate cost blackout periods
Provide a continuous evaluation of total project cost
Take corrective action to minimize cost increases
Provide an economic yardstick for measuring changes to the project basis

The program that industry has developed to meet these objectives is known by various names, including cost monitoring, cost trending, and deviation reporting. All of these programs are similar in nature and focus on evaluating the cost impact of changes to the existing estimate basis as these changes occur. A good cost-monitoring program meets its objectives by doing the following:

Keeping management continuously informed of the probable project cost, thus permitting early cancellation if the cost becomes uneconomical.
Focusing attention on cost increases and their causes in time for corrective action. This frequently results in improved economical designs.
Providing an economic analysis of each individual change before it is approved, thus giving management a logical basis for decision making.

THE FIRST COST BLACKOUT PERIOD

Let us go back and review project development during the first cost blackout period (i.e., between the preliminary and semidetailed estimates). It is during this stage that the following basic requirements of the project are established:

Plant size (capacity) is determined.
Number of products and product quality specifications are fixed.
The preferred process is selected and rough heat and material balances calculated.

Utility requirements are established and sources of utilities agreed upon (i.e., purchased or self-generated electrical power).

An exact plant site is selected. Previous planning may have considered a plant in the Chicago area; now a specific plot of land is acquired.

Pollution control requirements are established.

Approximate off-site requirements such as the following are determined:

Required maintenance and warehouse facilities.

Raw and finished product storage capacity.

Number of operating staff requiring office space and parking area.

These are the so-called "firm requirements" or design basis and, once established, set the foundation for the entire project. Decisions made at this stage determine the theoretical minimum cost of the project. Once the design basis has been set, the minimum investment or cost floor is established; therefore, cost control while these requirements are set is essential. Of all the phases of project development, this phase offers the greatest opportunities for cost savings. At no other stage will there be so many options open to management.

THE SECOND BLACKOUT PERIOD

The second blackout period—that which occurs between the semidetailed and the definitive estimates— is nearly as critical as the first. It is during this period that process specifications and flow sheets are finalized. Here are some of the things that are happening:

Process flow plans and specifications are optimized. These plans will show all major equipment specifying duties, operating temperatures and pressures, major pipelines, and operating controls.

The degree of plant automation is decided upon. This is the time to study computer vs. manual operation and remote-motor-operated valves vs. manual operation.

The height and diameter of major process vessels are calculated, and materials of construction are specified (i.e., carbon steel with a large corrosion allowance vs. alloy or stainless steel).

Soil borings are taken and foundation types determined (i.e., spread footings vs. pilings and, if pilings, what type).

Firm plans are made for providing utilities, i.e.:

Purchased vs. generated power

Once-through or recirculated cooling water

Air fin or shell and tube coolers

Method of waste disposal

Degrees of process flexibility and plant reliability are decided upon. The decisions to be made in this category include:

Philosophy with respect to spares: What pumps and compressors require spares? Should the cost of a spare be saved at the risk of plant shutdown? Can the plant operate at limited capacity for a brief period making 50% spare a reasonable risk?

Flexibility: Should the various units of a plant be interconnected in such a way as to permit one unit to be shutdown for repair while other units continue to operate? If so, the cost will be increased by bypasses, block valves, and intermediate product storage.

Reliability of utilities: Should evey electric substation be supplied with dual feeder cables, thus permitting uninterrupted operation while one cable is repaired? Should standby electric-generation equipment be installed to guard against accidental shutdown of power from the public utility?

The degree of preinvestment for future expansion is decided upon. A plant is designed to produce a specific quantity of product. However, the question always comes up, shall we incorporate some features now that will make future expansion easier? For example, should certain costly special equipment, such as fired heaters, be designed with spare capacity, thereby easing the problem of increasing production at some future date?

Firm requirements for office space, warehousing, and maintenance facilities are established.

Early project execution plans are formulated including the following:

Probable equipment delivery times are studied and a preliminary project schedule developed.

Feasibility studies are made as to how construction equipment and materials can be moved to the site. Temporary rail spurs and new roads may be needed, or a large ship-unloading dock might be required.

Studies are made to determine the availability and cost of labor at the construction site.

Economic forecasts are made covering escalation of wage rates and material costs.

All of the preceding items have a tremendous impact on the project cost. Without cost control, decisions made during this second blackout period can easily increase project costs by 50%, and this increase would not be discovered until completion of the definitive estimate. By that time, the owner would have spent a great deal of nonrecoverable money for engineering and long-delivery equipment, and he would be unable to reverse critical decisions and make cost reduction studies without halting engineering, thereby falling behind schedule. With cost control, each factor is identified and priced as it occurs, so that management can evluate economics as well as other pros and cons before making critical decisions.

THE PRELIMINARY ESTIMATE

A design basis must be established before a meaningful preliminary estimate can be developed, and good estimating practice requires that this basis be well documented. In process-plant work, it has been found by experience that neither the process planner nor the estimator can make an adequate preliminary estimate. The two persons (or groups) should work together as a team and be jointly responsible for the preliminary estimate and its documentation. This documentation is extremely important because it forms the basis of cost control through the conceptual- and early detailed-design phases.

CHANGES TO THE ESTIMATE BASIS

As we have seen, many changes to the basis of the preliminary estimate are made during conceptual engineering. For cost control purposes these changes should be classified in groups as follows:

1. Operational: This group includes all changes that are absolutely necessary for operation of the plant.
2. Safety: Changes necessary for safety of personnel or equipment are in this group.
3. Intangible benefits: Included here are changes that may be desirable but have only intangible benefits, such as elaborate landscaping or an impressive entrance gate.
4. Uncontrollable changes: Changes in this group are outside of management control, for example, revised pollution laws or an increase in state sales tax.
5. Economic: The majority of changes fall into this group, and it is here that cost control can be effective. The justification for an economic change must be in some form of tangible payout. This means that the change must do one of the following:
   a. Produce greater product quantity or improved product quality
   b. Reduce maintenance costs of labor or materials
   c. Reduce operating costs of labor, consumables, or utilities

YARDSTICK FOR JUSTIFYING ECONOMIC CHANGES

An economic yardstick should be developed for measuring changes classified as economic. The yardstick should be an accepted tool of management and rigorously applied to changes. In general, the yardstick should be a recognized method of calculating profitability that takes the time value of money into account. A number of these methods are in use and various companies prefer one method over another. One of the most common yardsticks is known as discounted cash flow. It is easy to apply and descriptions can be found in texts on economics. For small changes, many companies prefer to use simple payout time in years. This is calculated as

$$\text{Payout time in years} = \frac{\text{investment}}{\text{annual savings}}$$

The acceptable payout time is frequently taken as 2 years. With this yardstick, a small change that did not return the investment within 2 years would be rejected. The type of yardstick selected may vary, but it is important that the selected yardstick be impartially applied and that the savings be real.

The pertinent aspects of a change can be summarized as follows:

Description
Cost impact: the required incremental investment
Schedule impact (i.e., will the project completion date be changed?)
Classification
Justification: comparison with the yardstick

After these features have been determined the proposed change should be approved or rejected by the PM. This formal approval step should not be bypassed. It provides a self-discipline to an engineering organization. Since nobody really wants to be the author of a rejected change proposal, uneconomic changes are seldom proposed, and proposed changed are generally well thought-out in advance. The formal approval step acts as a powerful brake on the number and types of changes brought up for consideration.

HOW COST MONITORING IS CARRIED OUT

Cost monitoring during the conceptual phase can only be carried out by making the planners and designers understand that they are responsible for costs, and this can only be done by a strong, cost-conscious mangement. The program works as follows:

1.   The planner/designer who is completely familiar with and responsible for the basis of the original estimate identifies all changes and advises the CE.
2.   The PM similarly advises the CE of changes in project timing or other execution plans.
3.   Cost engineering keeps up to date on changes in material escalation or wage rates.
4.   All of these changes are fed to the CE, who develops the cost and schedule impact of each change and advises the cost control committee or PM.

The preceding points should be incorporated into a formal procedure for the guidance of all project personnel.

THE COST-MONITORING COMMITTEE

The planning of process plants and refineries requires technical expertise in a number of specialized fields, making it difficult for one individual to rule on the acceptance or rejection of changes; therefore, many managers make use of a cost-monitoring committee. Membership of such a committee for a large project might consist of the following persons:

The lead process planner or designer
The lead utility planner or designer
Technical specialists who consider only changes in their fields:
>   Instrument specialist
>   Electrical specialist
>   Environmental specialist
>   Piping and vessel specialist
>   Heat transfer specialist

The PM or coordinator
The project CE

This committee should meet at least once a month and perhaps more depending on the number of changes considered. The duties of the committee are as follows:

1.   Review flow plans to see what might be deleted. Process flow plans have a way of growing; pieces of equipment are continually added to improve the process or its control. Occasionally the addition of a piece of equipment or instrument in one area negates the need for a similar item included elsewhere on the flow plan. The committee must be alert and eliminate these "belt-and-suspender"-type items.

2.   Question any "firm" requirements that appear to cost an exorbitant amount. Occasionally the equipment to produce or handle a by-product looks low cost during early planning; however, during definitive planning it may be discovered that this equipment is costly. In such situations the committee should reevaluate the economics of producing the by-product.

3.   Define the economic yardstick to be applied to changes. This yardstick may not be the same for all projects, and it certainly will change over a period of several years. The committee should evaluate the yardstick as it applies to its specific project.

4.   Put out periodic (probably monthly) reports defining changes from the existing project basis and showing the cost and schedule impact of these changes.

Figure 2 shows the flow of information in a typical cost-monitoring program under the guidance of a cost control committee.

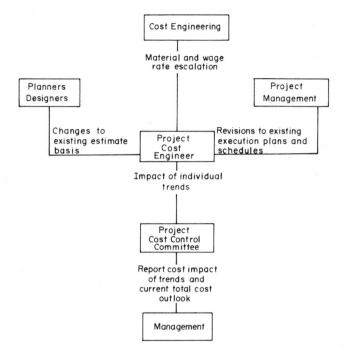

Figure 2   Cost trending: how it works.

THE COMMITTEE REPORT

The project cost-control report is the principal communication between the committee and the top company management; also, this report is given wide circulation among those working on the project. The report serves two functions:

1.   It informs company management as to the status of project development and the latest cost outlook.
2.   It provides project personnel (mostly specialists) with an overall view of the project.

It has been found that a report consisting of the following four sections provides the needed information:

1.   Narrative section covering
    a.   Schedule outlook, including dates for completion of conceptual engineering, detailed engineering, and construction.

    b. Highlights of progress made during the month, including discussion of any significant changes approved or rejected.

    c. Brief resumé of activities planned for the coming month.

2.   Project cost summary as shown in Table 1. This section should show the preliminary control estimate, all changes approved to date, and the revised cost outlook.

3.   A list of all changes approved during the month, as shown in Table 2. In addition to a brief description, the list should also show the cost and schedule impact, category, and payout for each change. Changes being prepared or under review on the date that the report is prepared should be shown under the heading "anticipated changes."

4.   A cost history of the project to date. This information can be shown graphically as in Fig. 3.

CONTROLLING THE COST OF CONCEPTUAL ENGINEERING

So far we have covered the monitoring of direct project costs (i.e., materials and labor); however, there is another dimension: the cost of conceptual engineering. This cost includes salaries and overheads of management and technical personnel

Table 1   Project Cost Summary, South Pole Refinery

| | Costs in $, thousands | | |
Description	Preliminary estimate	Approved changes	Revised est. 1/31/78
Material	13350	+ 320	13670
Labor	7200	+ 100	7300
Tankage S/C	2200		2200
Site prep S/C	1500	+ 200	1700
FLOH	4200	+ 60	4260
Engineering and fee	5000	+ 20	5020
Delivery costs	1550	+ 20	1570
Other overhead	720		720
Subtotal	35720	+ 720	36440
Contingency	7180	− 80	7100
Total	42900	+ 640	43540
Anticipated changes			− 50

        Cost engineer: John Kashflow

        Proj. manager: E. Scrooge

Table 2 Changes for Month of January, South Pole Refinery.

Item	Date	Description	Cost, $, thousands	Sched. months	Categ.	Payout years
Approved changes						
1	1/12/78	Increase size of H/F strip, tow.	+ 70	nil	Oper.	–
2	1/15/78	Increase site fill by 78000 cy.	+ 200	+ 0.5	Oper.	–
3	1/18/78	Add minicomputer	+ 400	nil	Econ.	2.5
4	1/22/78	Penguin crossings over pipeways (New legal requirement)	+ 50	nil	Uncontr.	–
Anticipated changes						
1	1/28/78	Reduction in loading rate	– 50	nil	Econ.	–
		Total anticipated changes	– 50	nil		
		Total approved changes	+ 720	+ 0.5		

Prepared by J. Kashflow
Approved by E. Scrooge

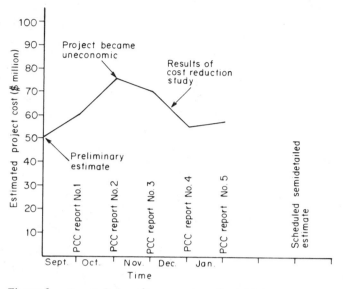

Figure 3   South Pole refinery cost outlook history.

engaged in performing conceptual engineering. By its nature, conceptual engineering demands a high concentration of specialist and management personnel who are also the highest paid persons in an engineering organization. At the time of writing, an average all-in-total cost for conceptual engineering averaged over $30 per manhour. Since a conceptual program can easily require from 50,000 to 150,000 manhours, one can see that cost control is important.

Budgets and schedules should be set up for the various planning and study activities. As soon as the initial concepts are established, major screening studies should be decided and a rough manhour budget approved by management. By the time the preliminary estimate stage is reached, firm engineering budgets and schedules for conceptual engineering should be in hand. Without a firm manhour budget and time schedule, conceptual engineering tends to drag on indefinitely piling cost upon cost. It is obvious that one should not spend 20,000 manhours on an alternative that could, at best, save only $20,000; however, this type of exercise is frequently carried out during the conceptual phase. The engineering budget should specify and include sufficient manhours for major economic design alternatives. A budget carefully planned by experienced persons becomes a valuable control tool, and the monthly report of the cost control committee should include forecasted conceptual-engineering manhours compared to the budget along with explanations of over- and underruns of either manhours or schedule.

Too much emphasis on staying within the engineering budget at this stage could lead to costly oversights. Justifiable overruns should be expected; the important thing is to make sure that they are justifiable. The object is to control conceptual-engineering manhours and funnel them into the study of design alternatives with real payout possibilities.

SUMMARY

The cost control committee's monthly report eliminates cost blackout periods by providing management with a continuous evaluation of total project cost. Each change to the project basis is itemized and priced in the report, and management can focus its corrective action efforts on the large significant changes. The economic yardstick provides a means of screening proposed changes and weeding out the less profitable. Also, a cost-trending program provides two intangible, but highly important, advantages:

1.   It instills a cost-conscious attitude in all project personnel.
2.   It keeps management informed as to total project cost, permitting cancellation at the earliest moment that economics become unfavorable.

14

COST CONTROL DURING DETAILED
ENGINEERING: INTRODUCTION
TO THE PROBLEM

FORD OR ROLLS ROYCE

As we have seen, the theoretical minimum cost of a project is set during conceptual engineering. If we call this minimum the cost floor, then we can say that detailed engineering sets the cost ceiling. It is during the detailed-engineering phase on a reimbursable project that we determine whether we get the Ford-type project visualized by the engineers during conceptual engineering or the Rolls Royce-type visualized by plant operators and managers. There is a natural conflict of interest between those persons charged with constructing a safe operating plant within a tight budget and those persons who will later have the responsibility of operating and maintaining the plant. This slight conflict of interest is healthy and the necessary compromises on both sides probably produce the most economical plant in the long run. The CE is a key figure in establishing the required check-and-balance system. His job—just as it was during conceptual engineering—is to identify, categorize, and price all changes to the estimate basis.

WHAT HAPPENS DURING DETAILED ENGINEERING?

Detailed engineering as used here actually means mechanical design of the plant. Conceptual engineering is largely confined to process engineering and culminates in a process flow diagram specifying equipment duties, sizes, capacities, and necessary control instruments, whereas the end product of detailed engineering is construction drawings specifying all details necessary for plant erection. Many owners do their own conceptual engineering, but for large projects they frequently look to contractors for the expertise needed in mechanical design.

During the period of detailed engineering the following activities are carried out:

Detailed piping and instrument (P&I) diagrams are made. These are extremely detailed process flow sheets showing all piping (both process and utility), valves, and instruments. The P&I diagrams along with narrative mechanical specifications provide the guidelines for all the remaining detailed engineering. These documents are second only to the design basis as an influence on project cost. Careless wording as to corrosion allowances or materials has the impact of cannon shots on project cost. *Careful review of the documents and comparison with the estimate basis is a cost-engineering must!*

Equipment pieces such as furnaces, heat exchangers, pumps, and compressors, are specified in detail and invitations to bid are sent out to fabricators and vendors. When bids are returned, they are compared as to both commercial terms and relative merits of the equipment offered, and a firm purchase order is placed for each piece of equipment.

Pipe routes are laid out—not only on paper, but frequently on an accurate scale model of the plant—and spool drawings are prepared in sufficient detail for the piping to be fabricated in a shop remote from the construction site.

Formal model reviews are held with owner/operator personnel, and if a model is not built, suitable drawings must be prepared enabling operators to review details of plant operation and maintenance. These reviews are the source of many changes and the CE should be closely involved.

Designs are made for miscellaneous bulk items, such as:

Foundations, paving, and sewers.

Structural steel for pipe supports, operating platforms, stairways, and ladders.

Electrical power supply for motors and lighting.

Instruments including sensing, measuring, and control elements plus signal transmission and air supply piping.

Buildings usually including control houses, offices, and warehouses.

Material takeoffs are made, bids requested from vendors, and purchase orders placed for the procurement of bulk materials.

Bids for major erection subcontracts are requested and awards made. If field erection is to be principally by subcontract, customary in many parts of Europe, the handling of subcontracts becomes a significant part of the total engineering effort. Minor subcontracts are frequently handled in the field by the construction manager outside of engineering responsibility. The award and administration of subcontracts is discussed in Chap. 19.

SHARPNESS IN THE DESIGN OF BULK MATERIALS

Sharpness in the design of piping and miscellaneous bulk materials can save a great deal of money. Frequently items are designed "fat" (i.e., are overdesigned). This can come about by inexperience or overconservatism on the part of the designer or through pressure to meet schedule and manhour budgets. The CE should be alert to overdesign. Examples picked up by the CEs on a recent project include the following:

Platforms and stairways designed too wide: Platforms are needed to provide access or working area and minimum requirements for specific locations should be set in consultation with experienced maintenance personnel. To save engineering effort, many design offices are inclined to use standards that frequently reflect maximum rather than minimum needs.

Code deflection requirements for steel buildings applied to pipe racks: Structural steel handbooks intended for buildings include deflection requirements small enough to prevent plaster from cracking. Beams selected from such a handbook may be larger than needed for pipe racks where greater deflections can be tolerated.

Advantages of combined footings overlooked: Considerable concrete can sometimes be saved by combining several pieces of equipment together on one foundation mat instead of using individual foundations. Frequently minor adjustments to the plot plan can make more combined footings possible, but since the persons who set the plot plan are not the ones who design foundations, these advantages are frequently overlooked. The CE is in a position to point out these possibilities.

Underground electrical cable incorrectly derated: The capacity of electrical cable is derated to allow for several cables in one trench. This became so routine in one office that it was extended to situations where there was only a single cable in a trench.

The preceding examples all occurred on a single project and are included here to illustrate the kinds of excesses that occur in the design of bulks. By use of bulk-sampling techniques, described in a later chapter, the CE roughly measures the sharpness of the bulk design.

A COST-CONSCIOUS ATMOSPHERE

During conceptual engineering the CE had to deal with at most two dozen engineers. It was possible to directly discuss costs with each person; however, during detailed engineering of a large project, he is confronted with 300 to 500 engineers, draftsmen, and specialists, all in a position to influence costs. The problem is how to make them all cost-conscious. A glance at Chap. 12, Fig. 2 will illustrate the magnitude of the problem. A cost-conscious atmosphere is the most effective element of a cost control program. The following are methods frequently used to help create this atmosphere:

Holding cost control meetings
Keeping each supervisor advised as to his budget and how his designs stand relative to that budget
Requiring an explanation for every design that exceeds the budget
Initiating a program of monetary rewards for exceptionally good cost-savings ideas

OWNER'S AND CONTRACTOR'S CEs

As previously mentioned, many owners turn to contractors for detailed engineering. The work may be done under any one of the several types of contracts discussed in Chap. 11. On most projects, whatever the type of contract, we will have two CEs—the owner's CE, resident in the contractor's office, and the contractor's CE. Both CEs have the same objective (i.e, to keep costs within the budget). Their responsibilities will vary depending on the type of contract: lump sum or reimbursable. A wholly lump sum job may require only a part-time owner's CE whose main responsibilities consist of evaluation of changes and verification of contractor progress used as a basis for partial payments.

On a reimbursable project the owner's CE is fully engaged in all aspects of the project. He works closely with the contractor's cost-engineering groups. The owner's and contractor's CEs should work together as a team with the mutual objective of realistically reporting the hard economic facts to their respective managements. On either type of job, lump sum or reimbursable, the contractor's CE has much the same function. Although on a reimbursable job the owner will be monitoring and sometimes criticizing his work, his responsibility will remain the same.

On both types of jobs some cost aspects are purely owner responsibilities. These usually include

Cost of owner's project team
Land
Taxes and duties
Currency splits (overseas jobs)
Financing
Operator training
Startup (contractor usually supplies assistance)
Spare-part review and warehousing

These duties plus the monitoring of changes and partial payments frequently require the full-time services of an owner's CE even on lump sum projects.

COST REPORTING vs. COST CONTROL

As mentioned in Chap. 11, there is a distinction between cost control and cost reporting. Referring to Table 1, cost reporting consists of the first three steps. Unfortunately, many PMs comply with only these first steps and consider that they are exercising cost control, whereas real control can only be obtained by carrying out the last three steps as follows:

1. Analyze: Find out what is wrong and why.
2. Take corrective action: Determine lower cost alternatives and study the pros and cons. There is always another way.
3. Check results of action taken: Follow up and determine that the action taken is achieving results. This follow-up is especially important in actions involving labor organization, supervision, or overtime.

CORRECTIVE ACTION

The CE's attitude is that corrective action to reduce a forecasted budget overrun is always possible. Of course, the ultimate corrective action is cancellation of part or all of the project. In the practical world, corrective actions are often worse than budget overruns. In these cases, budget overruns are accepted and included in the cost forecast. However, available lower cost alternatives should be considered for all budget overruns. These alternatives should be ruled out only after proven undesirable.

Corrective action is illustrated by the following examples from a recent project:

Comparison of expenditures for local field construction staff with the budget triggered an investigation that resulted in the reduction of 18 persons and a total savings of 157 workmonths.

Table 1   Cost Control vs. Cost Reporting

1.	Know what has to be done (control estimate)	
2.	Know what has been done (commitment record)	Cost reporting
3.	Know what remains to be done (forecasts)	
4.	Know how performance compares (analysis) to the budget	
5.	Minimize cost overruns (corrective action)	Cost control
6.	Check results of action taken (follow-up)	

Comparison of the budget vs. bids for surface condensers revealed that the budget basis of epoxy-lined water boxes was lower cost than the bid basis of Cu/Ni lining. Alternative bids verified that appreciable savings could be realized by going epoxy. In this case, the final decision favored Cu/Ni for maintenance reasons, but the budget comparison revealed a reasonable lower-first-cost route that might have been followed.

Sampling a group of computer-designed concrete footings showed that the volume of all small footings averaged between two and three times the budget estimate. Investigations revealed that all footings were being set 4 ft. below grade regardless of location or loading. Correction of this resulted in the savings of a considerable amount of concrete, forming, and excavation.

COST AREAS TO BE CONTROLLED

The cost areas to be controlled during detailed engineering can be classified into the following five categories:

1. Contractor engineering manhours and other office costs: The engineering manhour budget should be prepared early—say, within 2 months of the start of engineering. It should be broken down by functions, such as purchasing, design, drafting, and scheduling. As work progresses, expenditures in each category should be compared to the budget and realistic overruns and underruns forecast.

2. Equipment purchases: The estimated cost for each major piece of equipment should be listed in the budget and compared to actual cost. Since the original purchase order value seldom corresponds to the final cost of a sophisticated piece of equipment, it is usually necessary for the CE to keep a brief cost history of each equipment piece.

3. Detailed design of bulk materials: Sharpness in the design of bulks, such as concrete foundations and structural steel, can save a good deal of money. Design quantities should be compared to the budget at the earliest possible moment. Quick corrective action in this area can minimize what would otherwise become large overruns.

4. Procurement of bulk materials and awarding of installation subcontracts: Poor performance in the procurement of bulk materials or in the awarding of subcontracts for their installation can largely nullify the savings made by sharp design. Comprehensive procedures covering the procurement and handling of bulk materials in both the engineering office and field are required. If installation subcontracts are involved, special procedures are required to ensure that they are competitively priced and well written.

5. Project changes: This is the single most critical item in cost control. The budget and cost forecast must be kept up to date reflecting all changes in

the process design. If this is not done on a timely basis, project costs may be irretrievably out of control leaving no opportunity for effective corrective action.

In the following chapter we will consider methods and tools used by the CE and project management to effect control over the five preceding cost categories.

15

COST CONTROL DURING DETAILED ENGINEERING: CONTROL OF ENGINEERING COSTS

INTRODUCTION

As discussed in Chap. 14, control during detailed engineering falls into five distinct categories, and the first of these is the cost of engineering. The major contributor to cost is manhours, and before any degree of control can be exercised over engineering manhours, the CE must take the following three steps:

1. Develop a measuring stick (i.e., a realistic engineering manhour budget).
2. Establish a method of measuring progress (i.e., a scale for evaluating percent completion at any point in time).
3. Forecast manhours needed for completion (i.e., develop productivity relative to the budget and use it to predict the manhours required from any point in time to completion of the remaining work).

In the following paragraphs we will see how the CE accomplishes the first three steps.

FIRST STEP: THE ENGINEERING MANHOUR BUDGET

The budget is the CE's primary control tool. It is the base point from which subsequent forecasts are made. Engineering is broken down into a number of special activities, including design engineering, drafting, material procurement, scheduling, specification writing, and cost engineering. Each of these activities is normally headed by a supervisor, and separate manhour budgets should be established for each activity. The manhour budget multiplied by the average wage plus engineering expenses and computer cost become the total estimated engineering cost. Care must be exercised in setting the manhour budgets, and the usual procedure is to let each supervisor estimate the manhours required for his area of responsibility. The supervisor's estimates are compared with historical records from similar projects and other in-house data before they are accepted. This comparison is a must; otherwise, some supervisors will try to make their subsequent performance look good by setting fat budgets, and others may lack full knowledge of the scope of work and set lean budgets that will be overrun.

An owner engaged in reimbursable work should make a rough check of the contractor's engineering estimate. Many owners find that it pays to develop independent techniques for estimating engineering manhours. Using process flow plans as a basis, these techniques are usually based on the number of pipe lines and pieces of process equipment. After contractors have completed their estimates owners should check it or compare it to their own independent assessment. Points of variance should be discussed, and owners and contractors should agree on a realistic estimate representing an achievable goal. This estimate then becomes the engineering control budget.

To be useful as a control took, the engineering budget must be prepared early. It is a good idea to aim at completing the budget 2 months after the start of engineering. This gives the supervisors time to develop the full scope of work and still make a control tool available early enough for corrective action in troublesome areas.

SECOND STEP: A METHOD FOR MEASURING PROGRESS

To exercise effective control over engineering manhours, a manager must know three things:

1. Progress made to date
2. Number of manhours used
3. Calendar time used

Knowing these three things, one can forecast manhours and calendar time required to accomplish the uncompleted work. Also, it is possible to predict if existing manpower should be increased or decreased to complete the work on schedule or, conversely, how long the schedule must be shortened or extended to complete the work with the available manpower. Points 2 and 3 can be obtained directly from accounting records, but point 1 requires a special method of measurement.

The measuring method normally employed is to calculate a percent complete. This percent complete must include not only progress made on preparation of drawings but also purchasing, inspection, specialist design, and all other engineering activities. Most contractors have sophisticated methods for calculating the percent complete, and the following is the approach most generally used:

Engineering work is broken down into major categories and the contribution of
each of the various categories toward total percent complete is proportional to the percentage that the estimated manhours for that category
bears to the total estimated manhours for the project. For example, if
drafting is estimated to consume 30% of the total engineering manhours
and the drafting activity is 50% complete, then the contribution of drafting to the overall percent complete would equal 50% of 30%, or 15%.
The percent complete for each of the individual major activites is based on a
physical progress measurement. For example, drafting percent complete is
calculated by dividing the number of completed drawings by the number
of required drawings. Progress credit is given to partially completed
drawings according to a predetermined scale, such as:

	Drawing, % progress
General layout completed	30
Drawing complete with notes and dimensions	80
Drawing checked	95
Check corrections made and drawing issued for construction	100

Table 1 illustrates a method of calculating percent progress for the drafting function. In this example drafting is split into its various subcategories, and
physical progress in each subcategory is measured by counting completed
and partially completed drawings. The percent complete contribution of
each subcategory is proportional to the estimated manhours in that
category. The contribution of the entire drafting function toward total
engineering percent complete is, in turn, weighted and combined with
other functions in exactly the same manner as the subcategories of drafting are weighted and combined.

In practice, determining the engineering percent complete is a complicated
procedure involving many activites and methods of measuring physical progress.
Some organizations measure only what is known as production engineering
which includes only those activities leading to approved-for-construction drawings and material takeoffs. The production activities include writing mechanical
specifications, preparation of piping and instrument diagrams, mechanical
design, drafting, and material takeoff. These activites all produce tangible results
which can be measured. Progress is not measured on the so-called support activities including supervision, scheduling, cost control, procurement, expediting,

Table 1 South Pole Refinery Drafting Progress

| Drafting section | Drawings | | Physical % complete (3) | Manhours | | Weighted progress (6) |
	No. planned (1)	Equiv. no. complete (2)		Budget (4)	% Weighting (5)	
Flow plans	10	10	100	500	5	5
Layouts	10	10	100	500	5	5
Civil and structural	30	20	67	2100	22	15
Piping	60	20	33	4200	43	14
Mechanical	20	12	60	1400	15	9
Electrical	10	5	50	700	7	4
Instruments	5	1	20	300	3	1
Total				9700	100	53
						% progress

Note: Column 3 = column 2/column 1 X 100;
Column 6 = column 5 X column 3.

inspection, and engineering support during construction. These are considered to be overhead or time-related activities. This approach is analogous to construction where we have direct labor supported by indirect labor included in field labor overheads.

Every engineering organization has its methods and describing them in detail would require an entire book. However, if an owner is to place reliance in a contractor's method, the owner's CE must verify that the method includes the following points:

Progress of individual activities based on a realistic physical measure independent of the manhours consumed.

All engineering activities including supervision accounted for either as production engineering or as a support function.

Those using the methods, usually design supervisors, understand and use the method correctly. Many supervisors tend to slight this duty and give seat-of-the-pants completion percentages that can negate the best system in the world.

Provision for reweighting the various activities relative to one another if it is found that the initial weightings, based on the initial manhour estimate, are not realistic.

Owners will find that most contractors have detailed written procedures that they will make available to the owner when operating under reimbursable-type contracts.

THIRD STEP: FORECAST MANHOURS REQUIRED
TO COMPLETE THE REMAINING WORK

From any point in time the accounting record will tell us the manhours expended to date. No control can be exercised over these hours; they are history. If a manager is to exercise control over the unexpended manhours, he must know where to focus his attention (i.e., where the troublespots are and the magnitude of the trouble). The CE provides this information by

1. Forecasting the manhours required to complete the work in each major engineering category
2. Comparing the forecast category by category with the budget
3. Analyzing the underlying reasons in categories that show budget overruns
4. Comparing the forecasted manhours with workers available to determine if the calendar schedule can be met or if additional personnel is required.

Armed with the preceding information, the PM can exercise control and take corrective action.

From any given point in project development the forecast of engineering manhours required to complete the work is built up from detailed forecasts made by supervisors of each major engineering function. For example, the drafting supervisor requires each squad leader to forecast manhours required for his area of responsibility; he then adds these together and develops a total forecast for drafting. Unfortunately, this common-sense approach frequently yields the wrong answer. Few squad leaders and supervisors are skilled at forecasting required manpower, and the CE must check these forecasts for consistency and reasonableness. One such check is to determine manhours required for 1% progress as follows:

> Table 1 shows drafting to be 53% complete. If the accounting record shows 6,000 manhours used, then manhours for 1% complete = 6,000/53 = 113 and we can predict that manhours required for completion = (100 − 53) × 113 = 5,311. Actually we know that more than 5,311 manhours will be required because experience from many past projects tells us that manhours for the last 10% progress will be at least 10% greater than the project average; therefore, a rough forecast could be built up like this:

Used to date at 53% complete	= 6,000	manhours
Next 37% = 37 × 113 manhours	= 4,181	
Last 10% = 10 × 113 manhours × 1.1	= 1,243	
	11,424	
call	11,400	
Budget (from Table 1)	9,700	
Overrun	1,700	manhours

If the drafting supervisor's forecast is in this range, the CE will accept it, but if not, he must discuss the discrepancy with the drafting supervisor to verify if unusual circumstances exist. Working together and considering all available facts, the supervisor and the CE develop the most probable outlook for drafting manhours.

If, as in the case just cited, there is a forecasted budget overrun, the CE will help the drafting supervisor in the development of corrective action. Possible recommendations are as follows:

Delete the preparation of detailed drawings and isometrics for piping less than 2 in. in diameter. Pipe of this size can be run in the field without detailed drawings.

Do not detail the hookup for each individual instrument but show only typical hookups and let the field instrument foreman work out detailed variations.

Since these suggestions impact on field construction, the pros and cons would have to be discussed with the PM before the suggestions could be adopted.

The technique just described for forecasting drafting manhours can be applied to engineering as a whole. The composite percent complete combined with total manhours used will give the number of manhours consumed to date to accomplish 1% completion. By plotting 1% complete curves for a number of projects, it has been found that they fall into a pattern similar to Fig. 1, which has been taken from an actual project. The shapes of these curves are sensitive to the methods used for calculating percent complete, and a curve typical for one engineeering organization will not be typical for another. When properly prepared from historical data, these curves become useful forecasting tools.

Referring to Fig. 1, the budget line slopes upward reflecting change orders that are gradually increasing the budget. Note that the detailed-engineering estimate was not available until the 30% point in engineering completion. This is late, and the high manhours actually used at the 10 and 20% complete points gave early warning of a serious deficiency in the preliminary budget. The actual line has a typical shape, and by comparing with curves from past projects an early forecast as shown by the dotted line could have been made at the 20% complete

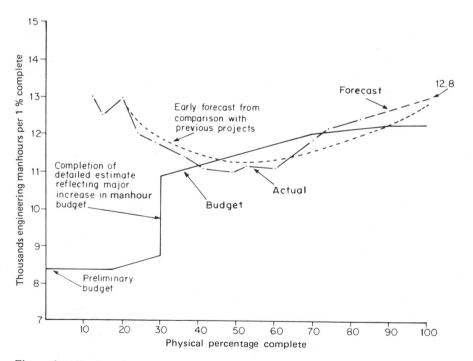

Figure 1   Engineering manhours for 1% completion.

point. Also, the large discrepancy between the preliminary budget and actual results shows that the engineering supervisors, who were reporting progress, had a good concept of the physical work to be accomplished that was not reflected in the preliminary budget but was belatedly recognized in the detailed estimate. There appears to be no good explanation as to why the detailed estimate was not developed earlier.

TOTAL ENGINEERING COST

We have discussed the three steps necessary for control of engineering manhours. Other elements that enter into engineering costs are the following:

Salaries and direct payroll burden
Fixed overheads including managment, accounting, and office space
Expenses including computer charges, travel costs, and reproduction services.

These three cost elements are discussed in detail in the following paragraphs.

Salaries and Direct Payroll Burden

Salaries are generally estimated and forecasted as averages. With several hundred people working on a project, the use of actual salaries for each person would be time consuming. The use of averaging concepts gives good results, and if historical curves, similar to Fig. 2, are plotted from past projects the problem is simplified. Figure 2 shows a typical pattern of variation in average engineering salaries over the life of a project. The average salary is usually highest at the start of a project when only managers, highly paid specialists, and designers are at work. The average reaches a low point at about 50% completion reflecting an influx of large numbers of draftsmen and technicians. There is usually an upward blip near the end of engineering reflecting higher paid follow-up personnel who are checking drawings, expediting vendors, and trouble-shooting field erection problems.

The salary pattern in Fig. 2 indicates in a general way what CEs should anticipate on their projects. They should, of course, plot their actual project and compare it to the pattern. Additional information that one would consider when making an actual forecast includes the average incremental wage rates paid during the previous several months, amount of overtime to be worked in the future, and salary categories of persons scheduled to remain on the project. On a large project CEs should plot a curve similar to Fig. 2, showing both cumulative and incremental rates, for each engineering activity or discipline as broken drawn in their budget and manpower forecast.

Payroll burden is usually estimated and forecast as a percentage of salaries. It covers the employers' cost of items such as social security, worker's accident liability insurance, medical insurance, vacation pay, pension plans, and any

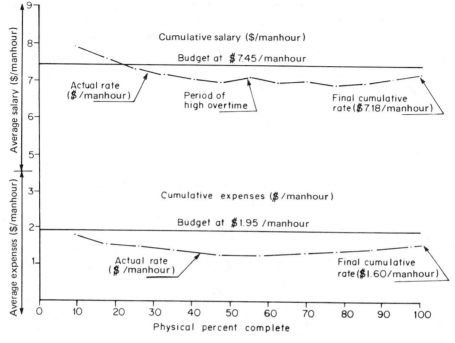

Figure 2   Engineering salaries and expenses.

other legal requirement or company policy that directly impacts the cost of retaining a worker on the payroll. Burden costs should be carefully investigated by estimators when preparing the control estimate. Thereafter, the CEs should occasionally monitor the actual cost to confirm that the estimator's percentage is correct. Normally burden costs are clear-cut and cause few problems for CEs; however, they can become a real headache in countries where employees receive extensive social benefits such as paid travel time, family medical care, and special allowances for school children. Also, CEs should bear in mind that a good construction safety record can usually be translated into substantially reduced premiums for accident liability insurance.

Fixed Overheads

Fixed overheads are prorated to projects on a dollars per manhour basis that is calculated by accounting methods and verified by auditors. These fixed costs include items such as office rental, utilities, janitorial services, and corporate management. The purpose of the proration is to allocate fixed costs to the various projects which are in-house at any particular time. The proration percentages are reexamined quarterly or semiannually, and most reimbursable contracts stipulate that any increase in the percentage allocation is subject to audit by the owner. There is little that the manager of an individual project can do about these fixed costs. Normally they are small, remain fairly stable, and do not represent a serious problem for CEs. However, the fixed overhead rate is a consideration that owners must take into account when selecting a contractor for a reimbursable project.

Engineering Expenses

In contrast to fixed overheads, engineering expenses are within the control of individual project managers. They can, for example, make more or fewer copies of specific project documents and make more or less use of computer techniques. These expenses normally average in the magnitude of 25% of salaries, but without careful control they can get out of hand and run much higher. Although there are a number of minor subaccounts that fall within the overall category of engineering expense, the ones that normally require close monitoring and management attention include the following:

1. Computer charges—items to watch here include:
    a. Too frequent issuance of extremely bulky complete reports when management type summaries may be sufficient.
    b. Continuing to run reports and distribute copies long after the need for them has passed. The issuing of many reports becomes routine.
    c. Running of too many design alternative cases when many of them can be ruled out by engineering judgement. Many engineers, especially owners' representatives, are inclined to say "run these cases anyway just so we can prove we looked at them."
    d. Retaining volumes of obsolete data in computer memory at a fixed cost per 1000 bits. Obsolete data should be taken out of memory, and if there is a remote chance that it could be needed again, stored on disk or tape.
2. Reproduction cost—items in this category that are often abused include:
    a. Distribution of copies of reports and letters to persons who do not need individual copies. Noncritical information can be distributed by means of daily pass-around reading files. Reports occasionally needed by several persons can be placed in a general file available to all.
    b. Failure to cut off the issuance of routine reports after the need for them has passed. Projects are dynamic and reports which are critical this month may be obsolete next month. For example, a weekly report on the status of design specifications may be important to the PM while the specifications are being written, but the report becomes useless after all specifications are complete.

3.   Travel cost—time consuming and expensive trips to vendors' shops, the construction site, and elsewhere, are frequently made when a well-written, letter and/or phone call could accomplish the same purpose. The PM needs to keep a tight rein on travel.

The various subaccounts making up engineering expenses are best monitored by means of trend or tracking curves. These plots show how cost varies with time. Figure 3 is an example of the subaccount for reproduction cost taken from an actual project. Trend curves are developed in the following steps:

1.   Establish an average calibration curve by plotting actual expenditures from two or three past projects. On these curves the percent of total costs spent should be plotted against the percent of engineering duration.
2.   Translate the average calibration curve into the budget and time frame for your particular project.
3.   Plot actual costs and compare to the calibration curve.

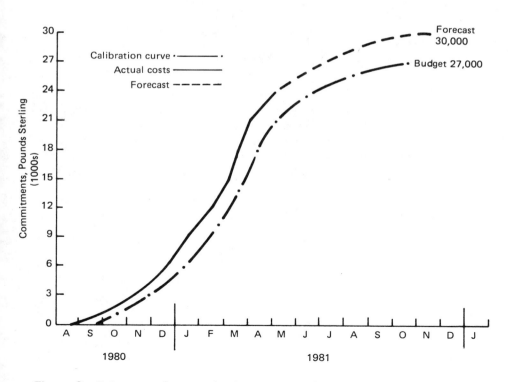

Figure 3   Subaccount for reproduction cost taken from an actual project.

4.   If actual costs track close to the calibration curve, no action need be taken.
5.   If actual costs begin to trend away from the calibration curve, either plus or minus, investigate to determine the reason. Note that a negative trend may mean costs are being mischarged into some other account.
6.   Look for possible corrective action and discuss with the PM.
7.   Forecast the total final cost.

A trend curve should be plotted for each of the main subaccounts included in engineering expense. Remember that trend curves do not control or forecast. They do, however, focus attention on the problem and measure the severity of deviation from the norm established by previous projects.

In Fig. 2 total engineering expenses are plotted as dollars per manhour vs. percent complete. This overall control tool is useful to owners and top management because it shows, at a glance, the actual engineering expenses compared to the budget and a trend towards overrun can be quickly spotted.

SUMMARY

In this chapter we covered the three critical steps which the CE must take to control engineering costs. These steps are:

1.   Develop a realistic engineering manhour budget.
2.   Set up a method to measure physical progress.
3.   On a monthly basis forecast the manhours required to complete the remaining work.

Remember, average engineering salaries do not remain constant throughout the duration of detailed engineering but vary according to fairly well defined historical patterns. Total engineering costs include not only salaries but also payroll burden, fixed overheads, and engineering expenses. Computer charges and reproduction costs are the two major expenses but other subaccounts are also important and should be monitored by trend or tracking curves.

16

COST CONTROL DURING DETAILED ENGINEERING: CONTROL OF EQUIPMENT PURCHASES

INTRODUCTION

The second category of cost control during detailed engineering is control of equipment purchases. The procurement of major process equipment consisting of towers, drums, reactors, pumps, compressors, and heat exchangers is one of the major activities during detailed engineering. Essential parameters and duties of the various equipment pieces are specified in the process flow plans and design specifications prepared during conceptual engineering. Preparation of mechanical specifications and procurement are the functions of detailed engineering. Some pieces of equipment are long delivery and must be ordered early in the project before the definitive estimate can be finalized. In these cases the estimator should make an early release of the estimate of the long-delivery equipment and have it in the hands of the project CE before vendor bids are opened. This procedure provides the CE with a base for comparing the bids and for determining at an early date whether or not he is faced with an overrun trend. Persistent overruns of early purchases is a storm signal that means the semidetailed estimate will probably be exceeded; it may also mean that the economy is moving into a high-escalation phase and that the economic justification for the project may be in jeopardy.

ADVANTAGES OF THE IN-HOUSE ESTIMATE

Equipment estimates prepared from information with the estimating organization (in-house) have advantages over the vendor quote-type estimates (out-house) as tools for cost control.

1. The in-house estimate can be prepared quickly and made available when needed, whereas out-house quotations frequently require a waiting period.
2. The in-house estimate can be related to a base time period, thus providing a fixed basis for measuring escalation.
3. The in-house estimate can be related back to the semidetailed estimate, which is the control tool used in early detailed engineering; therefore, over- or underrun trends to the semidetailed estimate can be easily determined.

It is the in-house estimate—related directly to the semidetailed estimate—that permits the CE to identify early trends in equipment purchasing.

COST HISTORY CARDS

Controlling and forecasting the costs related to equipment purchases are best accomplished by means of cost history cards, as illustrated in Fig. 1. Using budget data, a card is set up for each major piece of equipment. The budget is the first cost forecast, and as better information becomes available from mechanical design and vendor bids, the forecast is revised.

Figure 1 is the cost history card for a tower designated as T-101. The first entry on the card is budget information, including the estimated weight and wall thickness of the tower as well as the budgeted cost. The second entry reflecting the completed mechanical design specifies a wall thickness greater than that budgeted, and using this information the CE has increased his cost forecast by $1,100. Subsequently vendor bids were received, a bid tabulation prepared, and the low bid, $21,800, entered on the card. Also, a change authorization for clips and brackets was approved, adding $900 to both the budget and forecast.

Note that the cost outlook includes a $400 developmental allowance in addition to the low bid. Process equipment is seldom actually delivered at the low-bid price. There is always design development taking place between bidding and equipment delivery, and vendors are asked to make small additions and modifications to the original piece of equipment. The developmental allowance is intended to cover the cost of these additions. On the average, developmental allowance varies from 2% for towers and heat exchangers to 5% for compressors and drivers. If unused, developmental allowance can be progressively reduced on a time scale and finally adjusted to zero when the equipment arrives on site.

Many engineering contractors have replaced cost history cards with a computer printout. The printout contains the same information as the cards, and it is maintained up to date by preparing input sheets whenever new information is obtained. Although preparing input sheets takes as long as updating cards by hand, the computer offers the advantage of instant summarization plus the

		Quantity			Cost ($)			
Source	Unit	Estimate	Forecast	Over + under −	Estimate	Forecast	Over + under −	Remarks
Detailed estimate					22,000	22,000		
Diameter	ft.	5	5					
Ht. w/skirt	ft.	100	100					
Thickness	in.	1 3/16	1 3/16					
Weight	tons	25	25					
Outlook 10/2/72					22,000	22,000	-0-	
Drg 100101								
Thickness	in.	1 3/16	1 ¼	+ 5%				
Weight	tons	25	26	+ 5%		1,100		Added cost due to wt.
Outlook 11/2/72					22,000	23,100	+ 1,100	
Bid summary					21,800	21,800		
Change auth. #25					900	900		Clips and brackets
Development						400		
Outlook 12/1/72					22,900	23,100	+ 200	

Figure 1   Cost history card for tower T-101.

capability of sorting by developmental allowance. This is a convenient tool and simplifies the task of gradually deleting unused portions of the developmental allowance.

TRACKING EQUIPMENT ESCALATION

Since early 1974 escalation has become a serious factor in forecasting equipment costs. In periods of high economic activity vendors will not give firm bids. Under such conditions the buyer's tactic is to obtain a firm base bid to be modified by an escalation formula. The formula should be fair to both buyer and seller, and preferably it should be tied to some recognized cost index. The U.S. government's Wholesale Price Index can be used, but usually there are other indexes more directly applicable to specific types of equipment. Also, the vendor should be requested to invoice escalation separately from the base price. This practice facilitates invoice checking and avoids misunderstandings and bad feelings between vendor and buyer.

For sophisticated pieces of engineered equipment requiring appreciable engineering and fabrication time (i.e., large furnaces, boilers, compressors, etc.), a vendor's escalation may cover the following three cost elements: engineering, materials, and labor. In such cases a comprehensive escalation formula might take the following form:

$$E = P_B \left[\frac{0.1 E'}{E_B} + \frac{0.4 M'}{M_B} + \frac{0.5 L'}{L_B} \right]$$

where

E = total escalation to be invoiced
P_B = bid base price
E_B = value of engineering cost index at the time of the bid base price
E' = value of engineering cost index at the centroid of engineering cost
M_B = value of the materials cost index at the time of the bid base price
M' = value of the materials cost index at the time of materials procurement
L_B = value of the labor cost index at the time of the bid base price
L' = value of the labor cost index at the centroid of labor cost

The constants 0.1, 0.4, and 0.5 represent the proportional contribution of each cost element to the total bid base price. The given values are illustrative only and actual values must be negotiated in each situation. Of course, the total amount to be paid equals the bid base price (P_B) plus the escalation (E). In many situations engineering takes place immediately after placement of the purchase order; therefore, escalation on engineering is negligible and can be dropped from the formula.

When negotiating escalation formulae, buyers must take care that the time frame for calculating escalation for each of the three cost elements is carefully spelled out and understood by all parties. Vendors, of course, prefer to use the values of the cost indices at the time of equipment delivery which could easily be 12 to 18 months after the completion of engineering and procurement of materials. This position is untenable and should not be accepted by the buyer.

When applicable, forecasted escalation should be entered as an item on the cost history card. If escalation is tied to a formula the CE must project the formula forward to the time of invoicing escalation. The projection should be based on available economic indicators and knowledge of the business cycle. (Persons who are good at this should retire from cost engineering and make their fortunes in the stock market.) By progressively correcting his forecast to reflect "seen escalation," the CE can gradually close in on the final escalation.

In forecasting escalation the CE can rely on feedback from the purchasing department. Also, many large contractors and owners employ economists whose job it is to advise on escalation trends. However it is done, tracking escalation requires close follow-up, and in periods of high activity, estimated escalation must be frequently reviewed.

BID TABULATIONS

Competitive bids from speciality vendors are generally solicited for the purchase of process equipment. The bidding slate for an expensive piece of equipment must be carefully selected by someone with expert knowledge in the field. For example, if one wishes to buy a medium-duty chemical pump, they should not solicit bids from fabricators that specialize only in heavy-duty API pumps.* Also, care must be taken to include the "hungry vendor" (i.e, the person whose shop is half empty frequently gives the lowest bids). Many engineers tend to solicit only three bids and think that they have competition when in fact they may have contacted only the three high bidders. Solicit bids from all qualified vendors; search for the hungry vendor and don't overlook bargains that may be available from foreign vendors.

After bids are received they are summarized on a bid tabulation form. If the equipment purchased is at all sophisticated, all bids will not be on exactly the same basis. For example, in the case of a compressor, one vendor may include intercoolers while another vendor does not. All bids are adjusted to the same basis on the bid tabulation form, and in the example just given, the cost of separately purchased intercoolers would be added to the bid that excluded them. Bid tabs are generally prepared by specialists in the field, and it is frequently necessary to go back to the various bidders to obtain additional information. All bids should be adjusted to reflect cost of the equipment delivered to site. If

*API refers to specifications prepared by the American Petroleum Institute.

there are differential erection costs, these too must be included on the bid tab along with any difference in freight, customs duties, or taxes. Completed bid tabs should be carefully reviewed and signed by both the responsible supervisory engineer and the CE. The CE should enter the budgeted amount on the bid tab and attach an explanation of over- or underruns greater than 10%. The fully completed and properly signed bid tab along with the recommended vendor should be submitted to the PM for approval to purchase. The formal approval steps should never be overlooked, as they force a self-discipline throughout the organization and reduce the probability of carelessly prepared bid tabs that might result in costly errors. Important points concerning bid tabulations are listed in Table 1.

INVESTIGATION AND POSSIBLE CORRECTIVE ACTION

If the low bid for a piece of equipment is 10% or more over the budget, the CE and the responsible speciality engineer should determine the reasons and recommend corrective action when possible. Items to look for when investigating overruns include the following:

More exotic materials than anticipated in the budget: Examples are stainless steel pump casings substituted for carbon steel and stainless steel liners included in vessels at points where corrosion might occur.

Increased capacity or size, which occasionally occurs due to conservatism on the part of a designer: The overall concern of a design engineer is that items

Table 1 Bid Tabulations

All bids should be conditioned (i.e., put on the same basis).

This basis should be total-cost-delivered to the construction site including freight, duty, taxes, etc.

If there is a difference in installation cost, it must be included. In this case the bid tab would show total installed cost.

The budget amount should show on every bid tab along with the CE's initials.

In cases of budget overrun, corrective action should be taken.

If corrective action is unsuccessful and a sizeable overrun remains, an explanation should be attached to the bid tab before approval to purchase is granted.

There should be a written office procedure to ensure that the above things happen.

---

within his sphere of responsibility be adequate to do the job under all conditions. This natural fear of specifying something inadequate frequently leads to designs that are more than adequate (i.e., "overdesign").

Poor bidding slate, the greatest offender of all: People tend to favor equipment with which they are familiar; frequently owners specifiy that they will accept equipment from only one or two favorite suppliers. This can be a costly attitude and should be discouraged by CEs and cost-conscious PMs.

Unnecessary requirements inadvertently written into the specifications: The writing of specifications is a time-consuming and tedious task, and the common practice is to copy large portions of a new specification from old specifications. This sometimes leads to inadvertently including a requirement to overcome a specific condition that existed on the past job but does not exist on the present one.

There are hundreds of reasons for budget overruns and no list could be complete. The preceding list is intended only to give some ideas as to where to start looking.

When investigating overruns it is important not to admit the possibility of a low estimate until all avenues of corrective action have been investigated and rejected. It is a psychological fact that once a low budget is admitted, all efforts for corrective action cease. If a budget estimate is low, proper corrective action may minimize the overrun and this is the CE's objective. The tactic is to maintain the budget as an achievable objective until absolutely proved otherwise.

EQUIPMENT TRENDING

Equipment trending is a tool for early spotting of rapid inflationary trends. During the period of equipment purchasing the CE should keep a graph similar to Fig. 2. Here we have the ratio: actual cost divided by the budget, plotted against the percentage of equipment budget-committed. The solid line represents the cumulative ratio and the dashed line the ratio for each month. Given that the budget is a realistic objective and includes normal escalation, the monthly trend line will quickly indicate impending problems. Working with Fig. 2, which is from an actual project, the CE raised the alarm in April that the budget was in serious danger, and by May management was in position to consider the economic advisability of project postponement or cancellation.

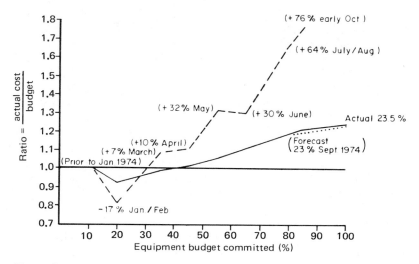

Figure 2   Equipment purchase.

SUMMARY

Complete and up-to-date cost history cards are essential for monitoring and controlling equipment purchases. Since major pieces of process equipment are usually custom fabricated, corrective action in overrun situations is frequently possible. Sometimes deletion of unessential requirements from the specs results in appreciable cost reductions. In periods of rapidly fluctuating escalation, money can often be saved by agreeing to an escalation formula. Insisting that vendors quote fixed prices forces them to be pessimistic and reflect high escalation rates in their bids.

The CE should enter the budget amount and initial all bid tabulations. In cases where a budget over or underrun exceeds, the CE and the specialty engineer should attach an explanation to the bid tab fefore seeking the PM's approval.

17

COST CONTROL DURING DETAILED ENGINEERING: CONTROL OF BULK MATERIALS

BULK MATERIALS

It is in the area of bulk materials, the third category of cost control during detailed engineering, that sharpness of engineering design enters the picture, and this is where a good engineering organization can save a great deal of money for the owner. In a previous chapter we discussed examples of overdesign that the CE should be alert for. Also, we discussed methods of estimating bulk quantities, which consist of

Concrete, rebar, and form lumber
Electrical cable, wire, and conduit
Instrument wire, tubing, and supports
Structural steel, ladders, and platforms
Paint and primer
Insulation
Piping

In this chapter we will discuss how the budget estimate can be used as a tool for forecasting and controlling both the quantity and cost of bulks. First, we will consider quantity control—the factor under the control of the designer. Second, we will discuss cost—the factor under control of the purchasing agent.

SAMPLING: A QUANTITY CONTROL TECHNIQUE

Sampling is the name given to the technique of selecting the first few completed drawings and comparing designed quantities to budgeted quantities. In the case of concrete, one selects the first few foundation drawings released from the drafting room and compares the designed volume of concrete to the detailed budget. Similarly, one compares piping lengths, electrical conduit lengths, power cable runs, and structural steel tonnages with their respective budget estimates.

If the first sampling shows a definite trend toward over- or underruns, the CE must investigate immediately and determine the reasons. In the case of concrete foundations, reasons for an overrun in volume might include the following:

Minimum depth of footings set well below that predicted in budget
Design strength of concrete less than the budget basis
Earthquake or wind loading considerations included in the design but not in the budget
Actual soil-bearing capacity less than the budget basis

As an illustration of concrete sampling consider Table 1 (p. 193) which summarizes a group of concrete samples from an actual project. In Table 1 we see that the six samples represent 5.5% of total expected project requirements; also, we note that collectively the samples reflect a 19% underrun in concrete volume and a 9% overrun in reinforcing bar. Since rebar normally represents only 25 to 33% of the total cost of foundation concrete materials, the sample indicates a definite financial savings and, at first glance, one might forecast that the remainig 94.5% of concrete yet to be designed will refect the same savings. However, before making such a forecast it would be advisable to determine the reasons behind the apparent cost underrun. In this case, analysis revealed the following:

Exchangers: The design underrun was entirely in foundations for air fin coolers E-807 and 857. Shell and tube exchanger foundations were close to budget quantities.
Six small reactors: The overrun was due to setting the bottom of the foundations of R-801 and 851 A & B at 4 ft. below grade. The estimate assumed an approximate depth of 2 ft.
Combinations footings: The estimate assumed an individual foundation for each piece of equipment; however, the designer saved an appreciable amount of concrete by placing a mat under several pieces of equipment: for example, the foundation for tower T-801 was used to support support drum D-804 and exchangers E-806 and 810. This type of construction requires more reinforcing and accounted for the overall overrun in reinforcing bar.
The allowable soil bearing had been changed from 3,000 psf used in the estimate to 5,000 psf used for design. This contributed to an underrun for all foundations, especially the tower and furnace foundations, which were without above-ground pedestals.

Table 1 Summary of Foundation Samples (5½% of Expected Project Requirements)[a]

Foundations for	Total concrete Cy			Total Rebar lb/100		
	Estimate	Drawings[b]	Over, + Under, −	Estimate	Drawings	Over, + Under, −
Exchangers	46	29	− 17	32.7	47.5	+ 14.8
Towers	293	184	− 109	351.5	159.1	− 192.4
Small reactors	52	74	+ 22	46.8	15.0	− 31.8
Drums	211	279	+ 68	165.2	217.0	+ 51.8
Furnace	61	32	− 29	122.0	74.0	− 48.0
Combinations of drums, exchangers and towers	538	372	− 166	434.0	745.9	+ 311.9
Overall	1201	970	− 231	1152.2	1258.5	+ 106.3
			19%			9%

[a] Estimated as individual footings.

[b] Includes 7% waste allowance.

The analysis shows that forecasting a 19% underrun for remaining concrete would not be correct unless one were certain that other opportunities existed for design of combination footings and air fin cooler foundations.

Forecasting, however, is not the only reason for early samples. An early sampling program frequently uncovers differences or discrepancies between the design basis and the control estimate basis. As discussed in a future paragraph, this may lead to corrective action in the form of adjustment/interpretation of specifications or to a recasting of the estimate to reflect actual conditions. As we shall see, when no fundamental basic differences exist, the recommended way of forecasting from samples is to plot each individual sample and cumulative trend on a chart similar to Fig. 1.

SAMPLING PIPING DESIGNS

In refineries and chemical plants piping is by far the most costly of the bulk materials, and piping erection will normally require about 25% of the total labor

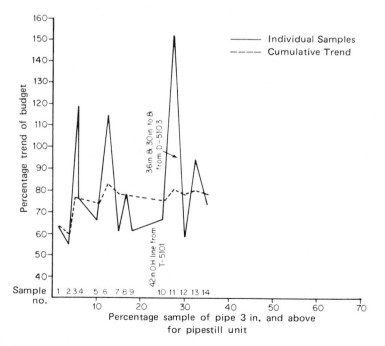

Figure 1   Pipe quantity trends obtained by sampling.

effort; therefore, piping merits careful sampling during the design stage. The detailed budget should list all major lines individually and show the size, pipe schedule, number of valves, estimated length, and number of fittings for each line. The size, schedule, and number of valves usually do not vary; the size and schedule are calculated according to well-defined rules and valves are shown on process flow designs. However, the length of pipe, number of welds, and number of fittings are within the control of the piping detailer, and these are the items that require sampling.

As soon as the initial group of piping layout drawings have been completed, the length and number of fittings should be compared to the budget on a line-by-line basis. Generally, a sampling of as few as 25 lines will indicate any significant over- or underdesign. Many leading contractors now build scale models of the plant, and the piping designer actually lays out lines on the model. This makes sampling much easier as the CE can count fittings and scale lengths directly from the model.

When using sampling as a forecasting tool, the results of each sample and the cumulative results of all samples to date should be plotted on a graph as illustrated in Fig. 1, which is a plot from an actual project. Although individual samples vary widely from the budget, one can see that a cumulative trend is developing at about the 20% complete stage, and the trend becomes stabilized at 80% of the budget by the time 30% of piping has been sampled. At this point a prudent CE might halt the sampling program and forecast a conservative average of 10% underrun for all piping. This forecast would then be held until the piping designers and material control engineers make their first detailed takeoff. Normally this takeoff would be made at about 70% design completion with 70% being actual and 30% estimated. The purpose of this takeoff is to firm up orders with piping suppliers and fabricators; however, it becomes a handy tool for the CE to use to adjust early forecasts made from sampling.

As with concrete, piping samples showing large deviations from the budget must be investigated. In Fig. 1 the large overrun of 30- and 36-in. pipe from drum D-5103 would certainly be investigated. There can be many reasons for overruns in individual lines; however, the following are frequently occurring reasons and can serve as starting points for investigation:

Overconservative allowance for thermal stress, frequently occurring when expansion loops are placed at arbitrary instead of designed intervals.
Poor equipment spacing due to overemphasis on accessibility for maintenance.
Excessive equipment elevation, sometimes caused by designer's desire to assure adequate NPSH for pumps.*
Careless pipe routing.

*NPSH is net positive suction head.

Arbitrary changes in elevation with change in direction, sometimes necessary in areas of congested piping, frequently extended into noncongested areas.

HOW TO USE SAMPLING RESULTS

As in any sampling program, care must be exercised in choosing representative samples. For example, if one carelessly chose piping samples that included only small sizes, say, 3 and 4 in. and neglected large sizes, the results could be completely erroneous. Similarly, a sample containing only alloy lines would not indicate results applicable to ordinary carbon steel piping.

When working with samples, deviations from the budget of less than 5% should be considered normal, and deviations between 5 and 10% bear watching and further sampling. Deviations greater than 10% should be investigated with corrective action in mind. Information obtained from sampling and subsequent investigation can be used for the following three purposes:

1. Corrective action: Sampling information frequently leads to relaxation of design assumptions, revisions to arbitrary office standards, relaxation of restrictive specification requirements, and clarification as to interpretation of specifications. Also, sampling occasionally picks up isolated design errors and misunderstandings among the various design disciplines.
2. Cost forecasting: Sampling provides early indications of quantity over- or underruns. Without sampling, the CE could not forecast until the completion of engineering takeoffs, and this would be very late indeed. One must be cautious in forecasting from samples. When forecasting from samples take the following precautions:
    a. Carefully select representative samples.
    b. Investigate underlying reasons for deviations.
    c. Plot sampling results and establish trends, as illustrated in Fig. 1.
3. Budget adjustment: It sometimes happens that sampling uncovers budget deficiencies or errors. When this happens,
    a. Make every effort to minimize the overrun.
    b. Adjust the designer's quantity budget by means of a change order. Put the budget on a realistic basis reflecting actual project conditions.
    c. Advise all concerned of the additional expenditures and reasons for the budget deficiency. Management is never pleased when advised of increased expenditures, but the bad news will be much better received during the early sampling stage than later at the final takeoff stage when it is too late for corrective or minimizing action.

MATERIAL TAKEOFFS

As mentioned previously, designers and material takeoff (MTO) engineers usually make bulk takeoffs at fixed stages of design completion, say, 20, 70, and

95%. These takeoffs are necessary for obtaining approximate quantities suitable for the early ordering of bulk materials. The 20% takeoff implies that 20% of the takeoff quantity is from completed drawings and that the other 80% is estimated or prorated from past projects. Usually this early takeoff is not as accurate as a detailed estimate; however, many organizations lacking sophisticated estimating techniques are forced to use the early takeoff as basis for the budget estimate. Needless to say, this is not a recommended practice. However, a takeoff made between the 60 and 75% completion points is usually reliable and the CE can use it to replace his sampling data.

If sampling information differs widely from the takeoff, the CE should do some reconciliation. He must be sure that

The MTO reflects the most recent drawings and changes
The takeoff methods used will produce reliable results
The personnel involved are experienced and competent

Taking off materials from an engineering drawing is a tedious, difficult task that is often poorly performed. The CE must be constantly alert and question conflicting data.

UNIT COSTS

As we know, the cost of bulk materials is made up of two parts: quantity and unit price. To save overall project time, it is usually necessary to go out for bids and establish unit prices before final design quantities are known. The usual practice is to predict quantities from budget or sampling data and then ask for unit price quotations. Occasionally vendors will state that their bid prices are only good under the condition that final quantities be within some range, say, ± 10% of the predicted quantities. When this happens, the probability of quantity variation must be evaluated on the bid tabulation.

In general bid tabulations (tabs) for bulks are treated similarly to bid tabs for equipment. In the case of bulks, the bid unit price must be multiplied by the forecasted quantity; this must be done for forecasting purposes, and also it highlights the impact of small differences in unit price. For example, a difference of 5 cents per sack of cement may not seem important until one multiples by the, say, 100,000 sacks required. Frequently the cost of freight becomes a significant factor when purchasing bulks. When reviewing bid tabs for bulks, the CE must be alert and verify that freight differential costs have been properly considered. This is especially true for large-size pipe and vendor shop-fabricated pipe. When comparing bids, always compare the cost of the material delivered to the job site.

FORECASTING ESCALATION OF BULK MATERIALS

Forecasting escalation for bulk materials is approached in the same manner as for equipment. Price at the time of shipment, which is the vendor's favorite approach to escalation, should be resisted by customers. As in the case of equipment, the buyer should negotiate for a firm base bid to be escalated by a formula, and the escalation should be invoiced separately. If price at time of shipment is the best term that can be negotiated, then the buyer should insist that the time of shipment be defined as the date that the raw material arrives in the vendor's shop and that this price be verified by an audit. If the customer has leverage and is willing to use it, he can normally negotiate an acceptable escalation clause; if not he should assume that the vendor is trying to secure an unfair advantage, and he is advised to take his business elsewhere.

Figure 2 is an example of an escalation tracking curve for prefabricated piping. This curve covers only materials that were price-at-time of shipment. Pipe fabrication labor had been covered by a separate escalation formula. The escalation projection started with the price at the time the purchase agreement was written ($1.41 per pound) and was projected forward to the centroid of the scheduled piping spool delivery date for each of the three construction areas. From the agreement date to mid-January, escalation was forecasted at 3.3% per

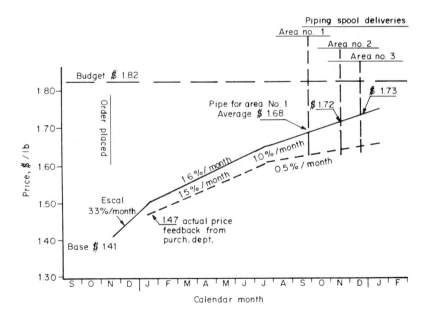

Figure 2   Material escalation, 1¼ in. chrome pipe.

month. After January it was expected to taper off to 1.6% per month and finally to 1.0% per month. In mid-January the contractor's purchasing department advised that the going rate was $1.47 per pound. This information indicated that the forecasted escalation of 3.3% per month was too high, and the CE made a second projection starting at the known base of $1.47 per pound. Charts of this type, while not perfect, at least provide some idea of what the total price will be and form a basis for predicting cash flow. Quite often these charts are made showing the anticipated escalation with a probable plus or minus range. This gives management and the financial people not only the probable total cost but also an upper and lower sensitivity range, and they can make contingency plans accordingly.

FINAL COMMODITY REVIEW

As the design of each bulk commodity approaches 95% completion, a final engineering commodity review should be held. The purpose of these reviews is to provide a firm cost forecast based on a near final material takeoff for each bulk commodity, (i.e., concrete, structural steel, piping, electrical, instruments, insulation, and paint). The data presented in these reviews are needed not only to forecast total material cost, but, more importantly, *to forecast construction labor manhours and to prepare construction schedules*. If the contractor's normal procedures do not include these reviews, owners, on reimbursable projects, should insist that the reviews be held.

It is suggested that the results of each review be presented at a formal meeting attended by both contractor's and owner's project managers. Each meeting should be chaired by the lead design engineer for the commodity under review, and it should be crystal clear that he is the person accountable for both quantity and cost as well as the technical aspects of all design work under his supervision. He is assisted by the material takeoff (MTO) and cost engineers who are, of course, responsible for the accuracy of the data which they develop.

It is important that each lead commodity design engineer understand, at the start of engineering, that he is responsible for the economy of his designs and that his work will be measured and compared to the control estimate. This direct accountability leads to tight, low cost designs; conversely, the absence of economic accountability permits sloppy (fat) and costly designs to go undetected.

The final commodity review wraps up each quantity/cost forecast. It includes allowances for designs not yet complete, field losses, and takeoff inaccuracies. Also, it forms a firm basis for construction scheduling and progress measurement. In the writer's experience, on both large and small projects, the forecasts made at commodity reviews have seldom required significant revisions throughout the remainder of the project, and the need for costly field material takeoffs has been minimized.

One final point: whenever possible, construction superintendents and field engineers should attend the commodity reviews falling in their areas of responsibility. This gives the field people an early firsthand overview of what they have to install and leads to improved relationships between field and office.

BULK MATERIAL CONTROL

The control of bulk materials is not a direct responsibility of the cost engineer, but it is an allied function. On some recent large projects the position of materials control engineer has been created. This person has the responsibility of coordinating all of the various functions involved with bulk materials.

The problems associated with bulks (i.e., quantity takeoffs, procurement, delivery, construction site warehousing, and retrieval as required) begin early in engineering and do not terminate until final turnover of the plant to operating personnel. Most engineering contractors now have sophisticated computer programs which aid in solving the problems associated with bulks, but the old problems of early identification of shortages, prompt procurement, rapid delivery, and site storage are still with us. These problems coupled with the contrasting need to minimize surpluses (materials left over after completion of construction) make efficient materials control a major concern of the project manager, and the potential loss/wastage of money make it a primary cost engineering concern.

A full discussion on bulk materials control is outside the scope of this book, but the CE should be concerned with the following:

1. Assure that a workable control system and good records exist. The workable system must include clear, concise procedures defining who does what and when. The procedures should explain in detail the control and reporting system to be used, either manual or computerized. Also, the procedures must cover the complex engineering office/construction site interface plus the receiving, storage and issuing of materials in the field. In short, we must always know:
    a.   What is required.
    b.   What has been bought.
    c.   Where it is located.
    d.   How much has been installed.
    e.   Anticipated shortfall still to be purchased, or surplus to be disposed of.
2. Confirm that good experienced people are involved in material takeoffs and that accountability for each activity clearly rests with a designated individual.
3. Make periodic appraisals to confirm that the procedures are followed and that the reported data is accurate. On reimbursable projects, a suggestion

in the ear of the owner's auditor will generally insure that adequate appraisals are made.

4. Confirm that proper allowances have been reflected in material takeoffs. Drawing takeoff is not a perfect science and allowances for site losses and breakage must be included. These allowances are made in the form of percentage increases (bumps) to the base drawing takeoff quantities. These percentage bumps should be based on statistical data from previous projects and not some engineer's blind guess.

5. Compare material takeoff (MTO) quantities to information gathered from sampling. If large differences should be found—say a significant budget underrun as indicated by samples turns into a large overrun after MTO— investigate and determine the reason. In the writer's experience several significant and potentially costly MTO errors have been uncovered in this manner. Remember, to err is human, to fail to reconcile differences between conflicting data can become costly negligence.

SUMMARY

Two factors enter into the cost of bulk materials: first, the quantity and second, the unit price. Both of these factors require careful management and control. Sampling is the technique used by the CE to get an early idea as to whether design quantities are trending over or under the budget. When spotted early, it's frequently possible to take corrective action that will minimize quantity overruns. Bid tabulations for bulks are handled much the same as for equipment pieces, except that in the case of bulks, unit prices msut be multiplied by estimated quantities in order to bring out significant cost differences between the bids.

Formal final commodity reviews are important control tools. These reviews are needed to firm up material cost forecasts, estimates of field man-hours, and schedules.

Bulk material control, while outside the direct responsibility of the CE, can have a decisive impact on project costs and it behooves the CE to make sure that this function is operating efficiently.

18

PROJECT CHANGES

CONTROL OF CHANGES: THE HEART OF COST CONTROL

Perhaps the most important aspect of the CE's job in the engineering office or on the construction site is the control of changes. No matter what else is done in cost control, if changes are not identified and estimated in a prompt, efficient manner, the project is out of control from the cost point of view. Change orders are number one priority for the CE.

Developmental changes to the project basis are a natural outgrowth of detailed engineering. The perfect design specification has never been written: there are always necessary and desirable changes to be made. As discussed in the estimating section of this book, the budget estimate must always include a contingency for changes. The CE tabulates the estimated cost of changes and sounds the alarm if the contingency is used up at a rate greater than the historical average established from previous similar projects.

THE BUDGET AND CHANGE ORDERS

Since the detailed budget estimate requires several months to complete and design engineering is taking place during that time, one can expect changes to be made to the design basis concurrently with development of the estimate. Since the estimator cannot estimate on a changing basis, it is necessary to establish a firm cut-off date for the estimate basis. Ideally the cut-off date corresponds to the first issue of the design specifications. Any process, project execution, or estimating change made after the cut-off date should be handled by means of a change authorization.

As change authorizations are made, estimated, and approved, they are added to the project budget; thus the current budget estimate always reflects the current scope of the project. The original budget estimate has a final cut-off date and reflects the project plan as it stood at that point in time; however, by the addition of change orders, the budget is kept alive and up to date, always reflecting the current project scope and execution plans.

If a project is executed on a lump sum basis, the change authorizations (CAs) represent hard cash to be paid to the contractor. From the owner's point of view, the same is true on a reimbursable project; a CA represents hard cash that must be paid. The difference lies in the fact that a lump sum contractor rarely will execute a change without a formal authorization; however, on a reimbursable job the contractor will add or delete items ar the owner's request without discussing costs, and without the self discipline of a CA system, an owner may add many things that increase the project cost and never realize that the budget has become obsolete. The owner without a strict CA system will frequently find that the project costs greatly exceed the budget and he will not be abel to identify the underlying reasons.

IMPORTANT ASPECTS OF CHANGE ORDERS

Important steps that the CE should take in handling project change orders (authorizations) are as follows:

Establish a written procedure and follow it!

A soon as a change is proposed, prepare an authorization including a preliminary estimate.

Prepare a detailed estimate as soon as possible; do not let a backlog pile up.

Determine the impact on the project schedule.

As soon as a change is estimated and the authorization approved add the cost into the current budget estimate, and incorporate the time impact into the current schedule.

If a CA is disapproved, add the cost of preparing the CA and estimate to the engineering budget.

Make it absolutely clear to the contractor that a verbal change given to a member of his staff by any member of the owner's staff is invalid unless backed up immediately in writing, preferably by means of an approved CA.

THE WRITTEN PROCEDURE

The written CA procedure should be precise and state exactly who does what. The responsibility for identifying a change should be clearly given to the area project engineer or designer, and the CE should be responsible for estimating the cost, determining schedule impact, and processing change orders. A high-level approval step should be included in the procedure; this is the brake that eliminates poorly thought-out and uneconomical changes. In no case is it wise to give approval authority to the person who originates the change.

Since it takes engineering time to estimate and prepare a change order, no CA should be completely cancelled. If the PM decides against the action recommended by a CA, the preparation of the CA remains as an added engineering cost, and the procedure should allow for this cost to be added to the engineering budget. Also, the procedure should provide for the sequential numbering of CAs and designate the CE as the person to control the numbering system and to keep a detailed log showing pertinent facts such as title, date, and estimated cost for each CA.

THE CA FORM

It is wise to give some thought to the preparation of the CA form. The approval form should be a single page, and backup documents can be attached as necessary to give a complete picture of the change. The following points should be included on the form:

CA number and date
Title and description
Brief justification
Originator
Category
Impact on project schedule
Estimated cost
Approval signiature

Figure 1 is an example of a form that has been used on several projects. Note that it is revision no. 1, detailed estimate. The original form was issued with a magnitude estimate only and showed a lump sum opposite total cost adjustment. After completion of the detailed estimate, revision no. 1 was issued showing an estimated cost breakdown and a revised total estimate. The total estimated cost, $13,300, was added to the project budget and to the cost outlook.

Since most changes involve both engineering and construction, the same form should be suitable for changes originating at the construction site as well as in the engineering office, and it is advisable to keep a separate file for each change since correspondence and backup documents tend to be bulky. The file should contain one copy of all information pertaining to the change.

CHANGE AUTHORIZATION NO. 041
SOUTH POLE REFINERY
REVISION NO. 1   DETAILED ESTIMATE

Date: 7/1/68

Description: This authorization covers a 6" bypass line around molten sulfur storage tank no. 2001. The bypass will connect to the fill line of tank no. 2002. This change will be included on a future change list to D.S. #67-32.

Justification: The bypass will allow sulfur rundown to be diverted around tank no. 2001 while ships are being loaded. This will permit the quantity of sulfur transferred to a ship to be measured by gaging the tank.

Originated by: South Pole Petroleum Corp.

Category: Operability or Safety X
Economic
Intangible Benefit
Uncontrollable

Estimated effect on target completion date: None

Estimated effect on project costs:

	Target cost
Engineering	$ 500
Materials	3500
Freight	-0-
Labor and burden	7300
Field labor overheads	2000
Fee	-0-
Total cost adjustment	$13,300

Project manager approval: E. Scrooge   Date: July 2, 1975

Figure 1   Change authorization.

ESTIMATES FOR CAs

Estimating the cost of changes must be given high priority. A large backlog of changes awaiting cost estimate is a sure sign of poor cost control. First it means that changes are approved without economic evaluation and second that nobody knows the estimated cost of the total project. Occasionally contractors attempt to estimate changes by using the pool of estimators in their central estimating department. This is poor practice; a full-time estimator completely familiar with all aspects of the project is needed on the project team.

A target timetable for the estimating of changes should be included in the change order procedure. An ideal time table is as follows:

Preliminary estimate: Within 2 days of identifying the change
Detailed estimate: Within 2 weeks after the preliminary estimate
Approval action: Within 1 week after the detailed estimate

When engineering on a large project is moving in high gear, it is not always possible to meet the preceding timetable. Normally the preliminary or magnitude estimate can be prepared within 2 days. The purpose of this estimate is to give the PM an idea of the total cost, say, within ± 25%, because he frequently must decide whether or not to authorize engineering to precede before the detailed estimate is ready. Frequently it takes more than 2 weeks to do the engineering necessary for preparation of a detailed estimate. In cases of this nature a target date for completion of the detailed estimate should be set and the PM advised.

CA estimates will be added to and become integral parts of the budget estimate; therefore, it is reasonable that they should be prepared to the same standard. For example, there is no point in making a CA estimate showing in detail the cost of nuts, bolts, and gaskets and then adding this estimate to a budget that has only a percentage allowance for piping accessories.

CAs are normally accepted at the estimated cost. It is seldom worthwhile to attempt to track the actual cost; however, there are situations where no estimating standards exist and there is genuine disagreement as to the probable cost of the change. In cases of this type, actual engineering, labor, and material costs can be accumulated in the project timekeeping records, but FLOH and engineering expenses must still be estimated, usually as a percentage of direct-labor and engineering salaries. Tracking the actual cost of individual CAs means laying out hard cash for timekeeping expenses, and for this reason it is advisable to minimize the use of this procedure.

For change order estimates the costs of construction FLOH and engineering expenses are usually added as percentages of direct-labor and engineering salaries. In the budget estimate these items are, of course, estimated in detail; however, for the purpose of change orders, the overhead accounts are usually reviewed in detail and reasonable percentages agreed upon to cover the accounts that are affected by small increases or decreases in the total scope of work.

CHANGE ORDERS PENDING APPROVAL

At any point in time there are some changes awaiting detailed estimates and others with detailed estimates but pending approval. The changes without detailed estimates should have preliminary or magnitude esitmates. The CE keeps a log of all change orders, and at any time he can tell the total value of changes both approved and pending. Normally the PM issues a monthly progress report, and one of the items included is a list of pending changes and their estimated value. With the previously mentioned 2-day limit for magnitude estimates, the CE can have a list of changes ready and evaluated within 2 days after the progress report cut-off date. The self-discipline of preparing a reasonable magnitude estimate for each pending change is an essential feature of cost control, and a backlog of changes without magnitude estimates or with wild guess-timates put down without logical backup is a sure danger signal that the project cost is getting out of hand.

WHO INITIATES CHANGE ORDERS?

All change orders (authorizations) should be initiated by the owner's PM. In virtually all cases the change is requested by the owner and the CA represent a written confirmation of his instructions. The contractor may point out the need for a change, and in some situations he may even refuse to proceed without a change, but he should not spend any time or money on a change without prior owner approval except in very unusual circumstances. The control of changes, in both lump sum and reimbursable projects, is an owner's responsibility, and if he wants cost control he must exercise that responsibility in a systematic and order-ly manner. An owner who permits members of his project team to verbally authorize changes by means of conversation with the contractor's draftsmen, designers, or construction foremen is violating a fundamental rule of cost control, and he is sure to encounter some unexpected cost overruns. A CE—either owner or contractor—cannot function under such conditions and should make this clear to the PM. Verbal authorization of changes, unless immediately backed up by a written CA, must be forbidden at every management level.

ESTIMATE ADJUSTMENTS

Estimate adjustments are used to correct obvious errors, omissions, or gross misjudgements reflected in the control estimate. In other words, estimate adjust-ments are used to correct estimating hacks. As we have seen, the control (budget) estimate is the yardstick by means of which actual performance is mea-sured; therefore, it must be kept realistic. The following are examples of estimate adjustments:

Correction of arithmetical errors

Addition of a piece of equipment, foundation, or pipeline that was inadvertently omitted from the budget estimate. There should be a budget for every item to be installed.

Correction of gross misjudgments or hacks such as estimating a concrete retain-ing wall to be one foot high when it is actually 10 feet high.

Estimate adjustments are not used to correct normal over- and underruns. For example, if a concrete foundation is estimated at 10 yd³ and the actual design is 8 or 12 yd³, no adjustment should be made. Likewise, no adjustment should be made for a pump estimated at $20,000 but actually costing $25,000. These are normal over/underrun to the budget and should be reflected in the monthly cost forecast.

Since estimate adjustments are additions or deletions to the budget estimate, they can be considered as a special category of change orders. However, since adjustments do not represent actual changes in the work, some project managers object to calling them change orders and prefer to use a separate form and numbering series.

BUDGET SHIFTS

When the control estimate (budget) is prepared, items which are known to be subcontract will, of course, be estimated as subcontract. However, during the course of the project, plans will change, and some items estimated to be subcontract may be executed by the prime contractor with direct hire labor. Other items estimated as direct hire may be changed to subcontract. These shifts in project execution plans are reflected in the control estimate by making *budget shifts*. A budget shift does not change the total estimated bottom line value, but it does shift funds between the various cost elements. For example, assume that because of a skilled labor shortage, a decision is made to subcontract the installation of a major 2500 yd³ foundation. The subcontractor is to supply all material, labor, and supervision required to complete the work including excavation and backfill. The control budget reflects the original execution plan which called for the prime contractor to furnish all materials and install the foundation with direct hire labor. A budget shift is required to transfer the estimated value of materials and labor from their respective accounts to subcontract. Also, since the subcontractor will be furnishing construction equipment, consumable supplies, and supervision, it is necessary to shift a proportional amount from the field labor overhead (FLOH) accounts to subcontracts. There will be no change in the total estimated cost—the increase in the subcontracts account will exactly balance the decreases in other accounts.

The same effect could have been achieved by means of a change order; however, it is best to reserve change orders for actual increases or decreases in the scope-of-work included in the control estimate. The preceding example did not change the amount of work to be done—it only revised the method of executing the work.

Budget shifts are necessary because our approach to cost control is a point by point comparison of actual plus forecasted expenditures with the control estimate (budget). If the comparison is to be meaningful, the control estimate must be kept alive and up to date reflecting the true scope-of-work and the current project execution plan. We have three tools for doing this—budget shifts, estimate adjustments, and change orders.

SUMMARY

In summary, the CA system is the cornerstone of cost control, and during the engineering phase, once a contractor has been given a change, he has no practical way of separating or differentiating this change from the remainder of his work. A piping change becomes part of an overall piping flow plan, drawing, and later, takeoff. Consequently, when the contractor makes monthly forecasts he is forced to include not only the original scope of work but also the effect of all changes initiated to date.

The CE's approach to cost control is a detailed one. This means an item-by-item comparison of performance with that predicted by the control estimate. Accordingly, it is essential that like items be compared. This can only be accomplished if changes, budget shifts, and estimate adjustments are processed promptly. Remember, if changes are ignored or verbally authorized and not evaluated, no matter what else is done, the job is running out of control.

In this chapter we discussed adjustment to the control budget by means of change orders, budget shifts, and estimate adjustments. These same items must, of course, be reflected in the cost forecast.

19

SUBCONTRACT DEVELOPMENT AND ADMINISTRATION

GENERAL

In Chap. 4 we discussed the preparation of estimates for subcontract labor and pointed out that subcontracting is the most common approach to field installation in Europe and Asia and, to a lesser degree, in the United States and Canada. There is a considerable difference between the approach and the techniques used to control subcontracts and that described earlier for controlling a direct-hire contract. The purpose of this chapter is to highlight these differences and outline the steps which the project management team and the cost control engineer must take to ensure that control is maintained throughout the execution of the project. The main emphasis will be on the three all-important "D's", definition, documentation, and discipline.

TYPES OF SUBCONTRACTS

Subcontract types can be defined according to their scope and/or their contractual basis. The subcontract scope can be for services (i.e., engineering and/or field supervision or other services), the supply of materials, fabrication, installation, or a combination of the above. Contractually, the more common approach is lump sum or unit price subcontracts. Reimbursable subcontracts are sometimes also used. These three types of contracts are discussed in more detail later.

SUBCONTRACT PHASES

Subcontracting and the control of subcontract costs can be split into two phases. The first phase is subcontract development and occurs during the detailed engineering phase. It involves the development of the subcontract strategy and plan, the preparation of the subcontract documents, contracting (i.e., invitation to bid, screening, selection), and finally, contract award.

The second phase is referred to as subcontract administration and involves the issuing of all contract documents, the monitoring of all changes (i.e., engineering, field and execution), the documentation of all verbal communications, the maintenance of a current log of the contract value, and the coordination of any claims.

Although, for major subcontracts, the main or prime contractor will usually assign a capable, full-time subcontracts administrator to coordinate and monitor all of the steps in both the subcontract development and the subcontract administration phases, the cost control engineer must be deeply involved, since the cost and schedule impact on the project is significant in both phases. In particular, the first phase, the development of the subcontract definition and the documentation, is often not given sufficient attention, with resulting commercial and execution problems developing later. Although the cost engineer will not be doing the administration, he needs to be fully aware of the steps involved and the potential cost and schedule impact to properly monitor costs in this critical area. Subcontract costs on a project which fully utilizes this approach can amount to 50% or more of the total project costs and many of these costs are set or established by the performance during the subcontract development stage. This is especially true if the subcontract is lump sum or unit price. Therefore, there is ample incentive for a major effort on the part of the cost engineer in this area.

SUBCONTRACT WORK DESCRIPTION

The single, most important element of the subcontract development phase is the subcontract work scope and definition. The invitation to bid and the contract documents should clearly spell out in considerable detail what the subcontractor will be required to do and, in particular, define the interface between his scope of work and responsibilities and that of others. Table 1 provides a summary checklist of the items which need to be considered in setting out the subcontract work description. Of critical importance is the degree of engineering completion, since detailed engineering is the main basis for determining the scope of work, the resource requirements, the schedule, and the cost. It is also important that the execution plan be established prior to letting of the subcontract. In particular, these considerations are important when the subcontract is lump sum or unit price, the most common contracting approaches. For the cost engineer and for the project, these two contracting approaches have major advantages over a reimbursable subcontract. Any form of fixed-price contract, such as lump sum or unit price, creates a strong discipline towards better work definition, both for

Table 1   Subcontract Work Description

Definition	
Specifications	Applicable Local Codes
Drawings	Basic Practices
Drawing Lists	Quanitities or Quantity Range
Interfaces with Existing Plant and Other Subcontractors	

Scope of Work	
Engineering	Inspection
Material Supply	Installation
Fabrication	Scaffolding
Delivery/Delivery Charges	Testing
Offloading	Commissioning
Storage	Spare Parts

the original contract scope and for later changes. It also leads to better planning and job execution and enables the cost engineer to forecast the final costs at an earlier date. This, of course, assumes that the work definition is good, the contract documents are sound, and the administration of the subcontract effective. If this is not the case, then major claims may be made by the subcontractor and this can create many problems in the area of cost control. This undesirable situation is discussed in detail later.

LUMP SUM VS. UNIT PRICE SUBCONTRACTS

The cost engineer must recognize that a unit price subcontract is a form of fixed price contract, similar to lump sum, but with some notable exceptions. In a unit price contract, the engineering is usually not completed and, consequently the final quantities to be installed are not specified. As a result, the schedule and the final costs are not firm. However, the risks for subcontractors and the incentive for them to exercise tight control is the same as in lump sum. Unfortunately, in most cases, because a lump sum commitment does not exist on a unit price contract, control on the part of the subcontractor in execution and on the main contractor and owner in administration and in changes is lax. Consequently, there is an incentive to convert a unit price subcontract to lump sum by either completing the engineering during the bidding period and convert at contract award or, alternatively, converting on a drawing by drawing basis, as each drawing is released for construction. The advantages of conversion to lump sum is that the final cost is known much earlier, it disciplines the documentation of engineering and engineering changes, and it can stop the "overworking" of the engineering phase. It also avoids a backlog of unpriced drawings and the associated cost reporting and cost control problems, especially at the final stages of the project. It reduces field staff, since all the pricing of the drawings will be done in the home office and, finally, because the subcontractor, the main contractor, and the owner are now looking at lump sum costs, psychologically this appears to create a much more cost-conscious atmosphere and tighter overall cost control. The cost engineer should ensure that every effort be made to either have lump sum contracts or unit price contracts, convertible to lump sum.

SUBCONTRACT COMMERCIAL TERMS

It is important to the cost engineer and to effective cost control that the commercial terms requested and included in the contract be comprehensive and cover most possible future changes or contract deviations. Table 2 lists some of the more important items which should be included. An effective job here can avoid later claims and makes for effective subcontract administration and lower overall subcontract costs. The basis for payment on a unit price contract should be on drawing quantities, determined in the home office and converted to lump sum, as discussed earlier. It should not be based on field measured quantities, since this occurs much later, incurs additional field costs to do the measuring and is unnecessary, since the drawing quantities are contractual and known before field work starts.

Table 2   Subcontract Commercial Terms

For lump sum contract, include:

    Unit prices, to be used for extra work
    Daywork rates for manhours and equipment hours
    Fixed cost per day, week, or month for delays
    Cost breakdown into areas subject to possible future changes (material costs, engineering, temporary facilities, mobilization, demobilization costs, etc.)
    Material supply or handling costs
    Overtime premium costs
    Bonus/penalty specifics, if applicable

For unit price contracts, include:

    All of above items, as applicable
    Effect on unit prices or large changes in scope
    Measurement for payment based on drawing quantities
    Option to convert to lump sum

BID TABULATIONS AND AWARDS

All subcontracts should be competitively bid and the bid tabulation procedure should be similar to that for purchase of major equipment discussed in Chap. 16. Many small subcontracts are awarded in the field and there is a strong tendency to bypass procedure and award the work to a contractor already on the site. The usual excuse for this is that bidding will take several weeks and the contractor on site can start work tomorrow. This is a poor excuse since normal work is scheduled well in advance and a decision to subcontract should not be delayed until the last minute. On reimbursable projects, the owner's CE should insist all subcontracts be competitively bid except on the unusual occasion when a real emergency exists.

SUBCONTRACT REGISTER

A subcontract register is an absolute must for cost control. This register comprises a complete list of all subcontractors and provides the following data on each:

Subcontract number
Name of subcontractor
Brief description of work to be performed
Date subcontract signed
Forecasted total value including an allowance for extras
Total amount paid to date
Percent complete to date
Anticipated completion date
Cost code to which subcontract is charged

SCHEDULE CONTROL

Tight and detailed schedule control of subcontracts is a must for the field cost engineer and scheduler. Slippage in a subcontractor's schedule generally impacts successive work which results in idle (waiting) time for other subcontractors or direct-hire labor—both of which are expensive. If a subcontract includes engineering, materials procurement, and field installation, every step must be monitored in detail. The field cost engineer and scheduler must make sure that *schedule reporting and monitoring authority are included in the terms of the subcontract.* If this is not done, some subcontractors will fall behind in engineering and material procurement, but they will conceal this information, report everything on schedule, and optimistically hope for a miracle to recover the lost time—never believe it, this is a recipe for budget overruns.

MONITORING AND CONTROL OF CHANGES

In Chap. 18, we discussed at length the monitoring and control of project changes. Many of the project changes also are changes to the subcontract. In addition, there are numerous other potential sources for changes to the subcontract, such as drawing revisions and modifications, field changes, changes to execution plans, and changed conditions, especially those related to the interface between the subcontractor and other subcontractors and with the existing plant. It is extremely important that all drawing changes, verbal instructions, and execution changes are documented, promptly and with a request for a cost and schedule evaluation by the subcontractor. To ensure no backlog develops, time limits on the response to these requests should be set and adhered to. There is no advantage to either party if there is a delay and unevaluated and/or unapproved changes are breeding grounds for later claims. This should be avoided. The most crucial time is the period immediately following contract award, when there is a need for the project manager, the subcontract administrator, and the cost engineer to establish a disciplined and structured approach. It is also important that a business-like atmosphere be maintained throughout the subcontract, with all changes identified, documented, and processed promptly—doing "favors" often only results in later claims.

Also, the prime contractor's procedures should make clear exactly who has the authority to authorize extra work by a subcontractor—preferably authority should be limited to the construction manager and the subcontracts administrator. All extras should be authorized in writing *with the agreed price stated,* and a copy would always be sent to the FCE. If a procedure of this type is not strictly adhered to, nobody will know the total value of outstanding subcontracts and the monthly cost forecast will be in error. It only takes a little relaxation of controls for subcontract costs to get out of hand. Remember—many subcontractors make their profit out of extras and claims—tight control is essential.

CLAIMS

It is difficult to address the subject of subcontracting and subcontract control without discussing claims. Unfortunately, most owners, main contractors, and the subcontractors seem to accept claims at the end of a project as an inevitable occurrence. However, as noted earlier, with proper definition, documentation, and discipline, there is no valid reason why most subcontracts cannot be closed out without claims. A large percentage of the claims are often, in reality, not claims but late requests by the subcontractor for approval of valid contractual changes. In other words, many of the "claims" result from inadequate administration of changes during execution (i.e., verbal requests never documented, changed conditions never recognized until towards the end of construction, sloppy handling or paperwork, and change order approvals) or, oftentimes, the result of the contractor and the subcontractor attempting to "tradeoff" one change for another. Most claims can be avoided by business-like but fair and equitable administration of the contract, and considering both the written contract and the intent and "spirit" of the contract.

There will, however, be situations where the cost engineer will become involved in major claim negotiations. Although the final settlement of the claim will be handled by the owner, the project manager, and the subcontracts administrator, the cost engineer will be called upon to contribute significantly by providing the cost history and evaluating the claim and reviewing the subcontractor's estimate of the claim. The claim from the subcontractor should always be submitted in writing, with detailed documentation of the reasons for the claim, the related facts, the contractual basis, if any, and a detailed estimate of the cost and schedule impact. If it is agreed that the claim is contractual, albeit late, it should be settled immediately, with any discussion centered around the cost estimate of the impact. If it is in a grey area (e.g., a "changed condition" such as a schedule delay caused by others) more discussion and work will be needed to determine the validity of the request and its real impact. If it is a "sympathy" claim (i.e., a request for financial assistance from subcontractors because they lost money despite their claim of good performance), the final decision as to whether to reimburse some or all of the loss will usually be made by upper management. In this case, their decision is influenced by the cost engineer's evaluation of the subcontractor costs vs. contract price.

It can be seen from our discussion on claims that the cost engineer has a vested interest in this area and must be deeply involved from the earliest stage. Where, despite the CE's efforts and those of the PM and the subcontract administrator, it is most probable that claims will have to be negotiated and paid, the cost engineer needs to include the best estimate of the claim settlement in the monthly forecast.

SUMMARY

We learned in this chapter that subcontracting is the most common approach to construction in Europe and the Far East and, to a lesser degree, in other locations. In areas where subcontracting is fully used, subcontract costs can amount to 50% or more of total project costs. Therefore, the CE must be deeply involved in this area.

There are two phases to the control of subcontracts—subcontract development, during the detailed engineering stage, and subcontract administration which occurs throughout project execution. The cost engineer has an interest in both.

The following are the key steps which must be taken to control subcontract costs:

Clearly define the subcontract scope or work definition.

Maximize lump sum subcontracts, converting unit price contracts, if necessary.

Include comprehensive commercial terms to cover anticipated and unanticipated future changes.

Document all verbal instructions and all changes, including changed conditions.

Evaluate and approve all changes as they occur—avoid backlogs.

Be business-like in administering the subcontract but be fair and equitable—consider intent in addition to the written word.

Never accept claims as inevitable, target for no claims at the end of the project.

20

COST CONTROL DURING CONSTRUCTION: THE FIRST THREE STEPS

INTRODUCTION

If a contract is fully reimbursable with a fixed construction fee, owner's cost control during construction is a must. Even in a target cost situation, where underrun of the target is shared between contractor and owner, the cost control effort needs some owner direction. In many construction organizations, the role of CE has been historically limited to cost reporting and estimating change orders. In many cases, the owner must take a strong hand if the benefits of cost control are to be realzed in the field. Most construction organizations are schedule-oriented and fail to consider the productivity debit that goes hand-in-hand with overtime or the high cost of having surplus construction equipment on hand in case it is needed to prevent a minor schedule delay. Also, construction organizations have a tendency to purchase the newest, most elaborate equipment available without really investigating lower cost machines to see if they will do the job equally well. When this philosophy is applied to cranes and bulldozers, the cost amounts to many thousands of dollars.

Prudent use of labor and construction equipment along with well-planned temporary facilites will save money for the owner. To get these, an owner needs a stong, knowledgeable field team to monitor the contractor's effort, and the CE is a key member.

THE OWNER'S FIELD COST ENGINEER

The owners' field cost engineers (FCEs) have the same duties as their counterpart in the design office. On many projects, they can do double duty, serving in the office during detailed engineering and relocating to the field for the construction phase.

One special duty of the owner's FCEs is to verify that contractor's schedule and cost forecasts are compatible. Construction cost forecasts are closely tied to schedule. If these functions are handled by separate groups in the contractor's organization, discrepancies are sure to occur and may result in unanticipated costs for the owner.

It is surprising how many times construction schedules are issued without correlating three principal elements:

1. Available workers on the payroll
2. Forecast of manhours required for completion
3. Number of working days available before the scheduled completion date

The owner's CEs are in an ideal position to review all three elements and verify that they are compatible.

ITEMS TO BE CONTROLLED IN THE FIELD

As in the design office, the budget is the FCEs' principal tool. His approach is a detailed comparison of budget cost vs. actual cost for field components. Major items to be controlled in the field are:

1. Direct labor: The biggest single cost and control problem on the construction site. Problems impacting labor cost include average wages, craftsman-to-helper ratios, overtime, weather, and productivity. Chapter 21 is devoted to methods of controlling and forecasting direct field labor.
2. Bulk materials: Purchasing, handling, storing, and controlling the use of bulk materials is of specific concern to the CE. Often warehousing procedures are lax and consumption records poor. Most construction supervisors concentrate on getting the job done and have little time left to worry about material cost and control.
3. Subcontracts: In the United States major contractors tend to directly employ their work force, but in many parts of the world the prime contractor subs out the work to small local firms. However, even in the United States, specialty subcontractors are always required, and their use can frequently reduce overall labor costs. CEs are concerned that bids are competitive and fairly evaluated. Also, they must forecast the total cost of each subcontract and then police the "extras" to keep costs in line with the estimate.
4. Field labor overheads (FLOH): When it comes to cost and control problems, FLOH ranks as a close second to direct labor. FLOH includes the following four subdivisions:
   a. Temporary construction
   b. Consumables
   c. Supervision and office expenses

d.   Construction tools and equipment

The CE's techniques for monitoring and forecasting these overhead accounts are discussed in Chap. 22.

5.   Other overheads: Every project has a group of miscellaneous overhead accounts that require the CE's attention. These accounts include items such as

a.   Material handling and storage

b.   Job cleanup

c.   Distribution of ice water

d.   Utilities used for construction

e.   Guards

It is impossible to make a complete list because every project has different requirements. These overhead accounts, frequently ignored by management, can become real dollar consumers requiring close attention from CEs.

THE FIRST THREE STEPS

CEs must take the following three important steps in the office before field construction gets underway:

1.   Set up a code of accounts to accumulate field costs.
2.   Recast their budget, making sure that there is a budget allocation for every cost code and that the field manhour budgets reflect the actual designed material quantities to be installed.
3.   Establish a procedure for measuring progress and calculating percent complete.

Only after taking these steps are CEs ready to function at the construction site.

The First Step: Code of Accounts

The first step for the FCE is to set up a code of accounts suitable for project cost control. As discussed in Chap. 8, most contractors have an established code covering the needed cost elements; however, the standard code usually needs modifications to suit the requirements of individual projects, and frequently owners have special accounting and tax requirements that must be met. In reimbursable contracts some owners force the contractor to adopt the owner's cost-reporting system. This is usually a costly mistake resulting in confusion and mistaken charges. The better approach is to use the contractor's code and make only the modifications necessary to satisfy any special tax requirements. A code of accounts should be tailored to meet two requirements:

1. Provide cost data necessary for cost control.
2. Provide a cost history satisfactory for tax requirements.

The code should be kept as simple as possible, and the control estimate, if not originally prepared by code, must be broken down to provide a budget for each cost code.

Many contractors use codes similar to that described in Chap. 8 and illustrated in Table 1. This seven-digit code provides the following breakdown, which is satisfactory for cost control:

The first two digits indicate the physical area of the work. Large projects are frequently divided into geographic areas for administrative and cost control purposes.

The third digit indicates the prime cost element—labor, material, subcontract, or other.

The fourth digit is used to divide the prime elements into significant subdivisions.

The fifth, sixth, and seventh digits are available for segregating construction details to the extent desired for management purposes.

Most contractors add a two digit prefix to identify the construction section or craft making the charge (i.e., pipe fabrication shop, carpenter shop, insulators, painters, etc.).

Referring to the example in Table 1, changing the third digit from a 1 to a 2 would change the translation from "labor for setting on-site formwork" to "material for on-site formwork."

In a lump sum situation the owner is only interested in getting enough cost information to satisfy tax requirements. In a reimbursable job the owner's FCE should be sure that the contractor's code provided the breakdowns necessary for cost control and the information for taxes. Also, the FCE must verify that there is a budget for every anticipated expenditure.

The Second Step: Recast the Budget

The second step for the FCE is to recast the manhour budget to reflect actual designed material quantities instead of the original estimated quantities. This recasted budget is usually referred to as the quantity-adjusted budget (QAB) and is calculated by multiplying the budget manhour rates by the actual quantities. For example, if the original budget for a foundation included 1,000 yd^3 of concrete at 10 manhours per yd^3 for a total of 10,000 manhours and the actual design quantity is 950 yd^3, then the QAB would equal 950 yd^3 times 10 manhours or 9,500 manhours indicating an underrun of 500 manhours. The QAB will represent either an over- or underrun of the actual manhour budget. It is not actually a new budget but rather a forecast reflecting designed material quantities to be installed at budget productivity, and it is used as a yardstick for measuring field achievement.

Table 1 Calculation of Percent Complete, South Pole Refinery

Cost code	Activity	QAB workhours, thousands	Weight percent	12/31/76 Physical % complete	Total proj.
0	Excavation	530	9	80	7.2
1	Concrete	640	11	70	7.7
2	Structural steel	150	3	50	1.5
3	Equipment and machinery	270	5	10	0.5
4	Electrical	360	6	10	0.6
5	Instruments	140	2	5	0.1
6	Piping	1920	33	15	5.0
7	Paint and insulation	510	9	0	0.0
8	Buildings	50	1	0	0.0
9	[a]Indirect	1160	21	35	7.4
	Total	5730	100		30.0

Prepared by J. Dollarspotter

[a]Indirect activities are normally part of field labor overhead (FLOH) and not reflected in job progress. The example represents an unusual situation where certain major temporary facilities were included at the owner's request.

The QAB is used:

To provide realistic manpower objectives for the field superintendents
To calculate labor productivity
To forecast actual costs
To prepare a realistic schedule
To weight the various construction activities for calculating overall percent
    complete

The QAB is the backbone of field cost control, and the preceding important uses will be explained and clarified as we further examine the duties of the FCE.

The final QAB cannot be established until the completion of design engineering, and this is usually well after the start of construction; therefore, the initial QAB must be based on forecasted quantities from sampling and early takeoffs. The FCE adjusts the QAB as better quantity data are supplied from the engineering office and, as we shall see, these adjustments have an impact on construction progress measurement. Since frequent adjustments to the QAB can cause fluctuations in progress measurement, it is advisable to use the actual budget as the QAB until such time as reasonably firm quantity information becomes available. Quantity data are developed in a code-by-code sequence as engineering progresses and final commodity reviews are completed (refer to Chap. 17); earthwork is normally first, followed by concrete and then structural steel. If the QAB is developed code by code in the same sequence as the final commodity reviews, the transition from budget to QAB becomes a smooth gradual process.

The Third Step: Calculation of Percent Complete

The third step for the FCE is to set up procedures for measuring construction progress. Progress is usually summarized and reported as percent complete. Most contractors have standard procedures, but these always must be modified to some extent to suit the peculiarities of specific projects. Unlike the QAB, the measuring system should be set up at the start of construction, preferably before the FCE leaves the home office.

The percent complete is a useful tool for project managment, and in many companies it is the only criterion of progress reported to upper management. It is frequently used as a rough schedule control tool. At the start of a project, scheduled percent complete is plotted against time, as shown in Fig. 1. As the job progresses actual percent complete is plotted and compared to the schedule. Graphs of this type can be plotted for the various subdivisions of a project or for the project as a whole. This type of graph is used by management to spot areas of possible difficulty. The graph is only used as an indicator; it takes a detailed analysis to discover if there is or is not a schedule problem.

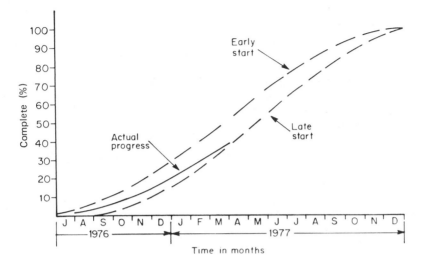

Figure 1   Construction progress.

As we shall see, the FCE uses the percent complete as a tool for measuring labor productivity and forecasting total direct field labor. Also, by relating to past projects, it can be used as a quick method of forecasting cash flow. There are many other uses for the percent complete, and the FCE must be sure that the method of calculation gives an accurate picture of job progress and that those using the method are using it properly.

METHODS OF CALCULATING THE PERCENT COMPLETE

In a previous chapter we discussed calculation of a percent complete for engineering, and the same approach used for engineering is applicable to construction, namely,

Construction work is broken down into major activities, such as earthwork, concrete, structural steel, and piping. The contribution of each of the various activities toward total percent complete is proportional to the percentage that the QAB manhours for that activity bears to the total QAB manhours for the project. For example, if the QAB for concrete represents 15% of the total QAB and concrete is 50% complete, then the contribution of concrete to the overall percent complete would equal 50% of 15%, or 7.5%.

The percent complete for each of the individual activities is based on a physical count in the field rather than the superintendent's judgment or the percentage of budgeted manhours used.

One must use care and judgement in selecting the basis for physical count. For example, the percent complete for concrete foundations would equal the cubic yards poured divided by the total cubic yards in the project if all foundations were similar; however, it would not be true if the project included many small complex foundations of 1 or 2 yd^3 each and say two large foundations of 400 yd^3 each. It would not be true because the manhours per cubic yard for small foundations is appreciably different than for large foundations. For the purpose of calculating a percent complete for concrete, one might divide the overall concrete activity into subactivities, such as

Large foundations over 5 yd^3 each
Pads and footings less than 5 yd^3 each
Walls, columns, and structures
Paving

Each of these subactivities would then receive a percentage weighting equal to its QAB divided by the total QAB for concrete, and the progress within each subactivity would be measured by physical count.

The physical count for such items as electrical wiring and instrument tubing is difficult. Some contractors calculate the total linear feet of wire to be pulled into conduit and then measure their daily progress by the amount of wire actually pulled. An easier method, and perhaps just as accurate, is to count the number of terminations made and report percent complete as the number of completed terminations divided by the total number of terminations.

Table 1 is a simplified example of a percent complete calculation. If we look at cost code 1, we see that concrete represents 11% of the total project and it is 70% complete, thus contributing 11% of 70%, or 7.7%, to the overall project completion of 30%. Since each of the major activities in Table 1 is divided into several subactivities, the entire process becomes a lengthy hand calculation. This is especially true if the project is divided into a number of geographic areas with a percent complete required for each area. Fortunately it can be easily computerized, and most large contractors have programs that require only an input of physical quantities completed during the reporting period.

BASIS FOR WEIGHTINGS

Up to this point, and in Table 1, we have used the QAB as the basis for calculating the relative weight of each of the various activities; however, many contractors prefer to use the current forecasted manhours. Use of the current forecasted manhours has the following two distinct disadvantages:

1.   The percent complete changes with every change in the forecasted man-hours for individual accounts. From Table 1 one can see that if QAB man-hours are replaced by forecasted manhours, any change in the forecast will change the weighting and therefore change the percent complete. A change of this type, completely independent of work accomplished in the field, confuses construction superintendents and tends to undermine the FCE's credibility.

2.   Use of the forecast tends to reward incompetent superintendents. As they require additional manhours to accomplish their work, their weighting increases and their portion of the total project appears to increase. Conversely the forecasted manhours and therefore the weightings for good superintendents are progressively decreased, and their apparent contribution to the overall project decreases.

Contractors who prefer to base the percent complete on the forecast argue that the QAB is often unrealistic. This objection is valid if the original budget unit rates (manhours per unit of work) are grossly too high or low. Budget unit rates must represent realistic achievable objectives; otherwise the QAB is invalid for all purposes. Another factor that can invalidate the QAB is failure to keep change orders up to date. Our whole premise of cost control requires that the budget be realistic and that it be kept current, and if this is done, the QAB provides a firm, fair basis for calculating percent complete.

INTERDEPENDENCE OF SCHEDULING AND MANHOUR FORECASTING

Manhour forecasting and job scheduling are both dependent on the percent complete, and all three must be compatible. Given a forecast of manhours needed for completion and the scheduled working days remaining, managers can quickly calculate the size of their required labor force and judge the necessity of scheduled overtime or additional hiring. Many contractors still separate scheduling and manpower planning from cost forecasting and control. This invariably leads to inconsistency and errors. It is obvious that size of the labor force, working days remaining in the schedule, and manhours required for completion must be compatible, yet for some unaccountable reason, this obvious point is frequently overlooked and the results are high overtime, frantic efforts to hire labor on short notice, schedule extension, and increased cost.

SUMMARY

In this chapter we have discussed the three fundamental steps that FCEs must take before they can forecast direct field labor. Two of these steps, the code of accounts and the procedure for measuring progress, should be accomplished before the start of construction. The third step, quantity-adjusted budget, must be handled on a code-by-code basis according to the engineering completion sequence and cannot be finalized until engineering is complete. In the next chapter we will see how these fundamental tools are used by FCEs.

21

COST CONTROL DURING CONSTRUCTION: FORECASTING DIRECT FIELD LABOR

INTRODUCTION

In construction terminology the expression "direct labor" is used to signify labor required for direct construction of the permanent facilities, and the expression "indirect labor" refers to labor required for temporary facilities, time-keeping, warehousing, and construction equipment repair. On a direct-hire, reimbursable job the CE's most difficult job is forecasting direct labor. For cost and schedule control, construction management requires forecasts of total direct labor updated on a monthly basis. Formerly, labor forecasts were frequently left to the field superintendents, who worked mainly by experience and crystal-ball gazing. We now have some tools that remove much of the guesswork from forecasting, but input from experienced supers should not be ignored. Each monthly forecast required the combined efforts of the supers with their experience and the FCEs with their tools. In the early stages of construction, when it is difficult to visualize the full scope of work, tools give better results than experience; however, in the final stages the supers' plans for completing the work and lay-off schedule give the best results.

As we all know, there are really two parts to labor costs: first, the number of manhours required, and second, the average wage or dollars per manhour. Since manhours is the more difficult of the two, we will discuss manhours first and average wage second.

FORECASTING TOOLS

The FCE's kit of tools for forecasting direct field manhours includes the following:

The control estimate (budget): This is the first forecast.

Material takeoffs: The budget manhour outlook is modified to reflect actual designed quantities. We call this the quantity-adjusted budget (QAB).

Productivity profiles: The QAB is modified to reflect actual construction experience.

Crew sizes and layoff schedules: These become important during the final stages of construction.

LABOR PRODUCTIVITY

A knowledge of how actual job productivity compares to that predicted in the control estimate (budget) is the key to forecasting final direct labor manhours. When preparing the control estimate, the estimator used a base productivity relative to the Gulf Coast, Chicago Area, or some standard location. However, once construction starts we are no longer interested in the Gulf Coast or any standard location. We accept the estimated productivity reflected in the QAB as our control base and set it equal to 1.0.

By using the QAB as a base, we can define actual productivity for any unit of work or for the job as a whole as being equal to the budgeted manhours divided by the actual manhours used to accomplish the work. For example, if the QAB for installing concrete was 800 manhours and 1000 manhours were actually used, productivity to the QAB base would equal 0.8.

The preceding calculation can only be made after the subject work has been completed and actual manhours are known. To exercise control and forecast we need to calculate productivity during the progress of the work. We can do this at any time by using the relationship: productivity equals the percent physical completion divided by the percent of QAB manhours used. In other words, if the concrete activity is 80% complete and 100% of the QAB manhours have been used, the productivity to the QAB base equals 0.80.

A third method of calculating productivity cannot be used for the job as a whole. It is applicable to activities with readily measurable quantities, such as cubic yards of concrete foundations and lineal feet of electrical conduit. This method uses the relationship that the productivity of any individual activity is equal to the budget unit manhour rate divided by the actual unit rate. For example, if the QAB unit rate for pouring concrete foundations is 8 manhours per cubic yard and the actual rate is 10 manhours per cubic yard, then productivity to the QAB base equals 0.8.

The three formulae for calculating productivity to the QAB base are summarized as follows:

1. $\text{Productivity} = \dfrac{\text{QAB}}{\text{actual manhours}}$

2. Productivity $= \dfrac{\% \text{ physical completion}}{\% \text{ of QAB manhours used}}$

3. Productivity $= \dfrac{\text{budget unit rate}}{\text{actual unit rate}}$

Experience shows that productivity does not remain constant throughout the life of a construction project but varies according to well-established curves or patterns called productivity profiles.

The shape of productivity profiles differs for each of the various activities, such as concreting, installing structural steel, and erecting pipe. These profiles are sensitive to methods used for calculating percent complete and may vary between contractors. Typically, productivity for any activity starts out poor, reaches a maximum between 30 and 80% completion, and then tapers off to final completion. A typical productivity profile with the theoretical curve derived from a previous project is shown in Fig. 1.

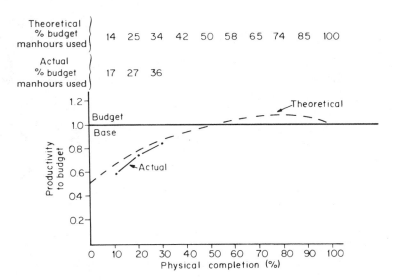

Figure 1 Above-ground piping labor productivity. Productivity to budget = % physical completion ÷ % budget manhours used. (Cost engineer, John Kashflow.)

HOW PRODUCTIVITY VARIES

Actual productivity profiles tend to be more irregular than the calibration curve in Fig. 1. These irregularities are caused by extraneous events, usually of short duration, including the following:

Weather: A month of exceptionally bad or good weather will cause the profile to deviate.
Length of working day: High overtime usually causes a drop in productivity.
Density of labor: When all crafts are working in the same area at the same time productivity tends to decline.
Supervision: Good or bad supervision has a big impact on productivity.
Union negotiations: Productivity frequently declines during a negotiating period.
Craft jurisdictional disputes: These disputes and work stoppages or slowdowns can cause big dips in productivity.

There are many other happenings that impact productivity, and when one analyzes the profiles and uses them as forecasting tools the reasons for irregularities must be kept in mind. Also, apparent fluctuations can be caused by shortcomings in the measurement of percent complete. Since no percent complete measurement is perfect, month-to-month peaks and valleys in productivity profiles are normal.

HOW CALIBRATION CURVES ARE DEVELOPED

The theoretical or calibration curve shown in Fig. 1 is for an underdeveloped nation or an area where skilled labor is not readily available and on-the-job training is required. In most areas of the United States the profile is much flatter, often starting at productivities of 0.85 or 0.90 and reaching a peak between 20 and 30% complete. The theoretical profile is established by plotting data from several similar past jobs where the same methods for calculating percent complete have been used.

Once the theoretical or calibration curve has been drawn, the theoretical percentage of budget manhours to achieve any given percent complete can be calculated as in the following example from Fig. 1:

1.   At 10% complete we read the theoretical productivity from Fig. 1 as 0.70.
2.   We then divide the percent complete (10%) by the productivity (70%) to obtain the theoretical percentage of budgeted manhours (14%).

What we have developed here is the theoretical rate of using up our QAB manhours. By plotting the actual productivity and comparing it to the theoretical, we have a tool for forecasting total manhours to project completion.

FORECASTING FROM PRODUCTIVITY PROFILES

Let us suppose that a construction project contains 120,000 QAB manhours and has a calibration profile similar to Fig. 1. What will be the forecasted overrun when 36% of the budgeted manhours has been used and the project is 30% complete? We calculate the actual productivity as:

$$\text{Actual productivity} = \frac{30}{36} = 0.83$$

From Fig. 1 we obtain the theoretical productivity at 30% completion as 0.88. We now determine the difference between theoretical and actual productivities as:

$$\% \text{ Difference in productivity} = [(0.88 - 0.83) \div 0.88] \times 100 = 5.7\%.$$

The forecasted overrun will be 5.7% of the budgeted manhours (i.e., 0.057 × 120,000 = 6,800 manhours).

OTHER USES OF PRODUCTIVITY PROFILES

The preceding example illustrates one use of the profiles. Another good use on a large project is to compare the effectiveness of area superintendents by plotting overall curves for the area supervised by each superintendent. This frequently lands FCEs into the center of controversy, but that's where they should be. It always stimulates supers with poor productivity to review their organization and either improve productivity or produce underlying reasons as to why it cannot be improved. With proper use this method never fails to make the entire organization manpower-conscious, which is one of the FCE's prime objectives.

Another important use of the profiles is to pinpoint trouble spots. If one plots period productivity on a monthly basis on the same chart with cumulative productivity, they can isolate adverse trends quickly. A month of rain or work slowdown will show up as a sharp drop in all of the charts; however, a sharp drop in one area or one craft indicates a problem and requires investigation. Later, we will discuss this use of the charts in greater depth.

TREND CURVES

Productivity profiles are useful tools for FCEs, but manhour trend charts give a more complete picture and are frequently used for presentations to company management. Trend charts are simply manhours plotted against percent complete, as illustrated in Fig. 2. The trend chart calibration (theoretical) curve is developed directly from the productivity profile. The calibration curve of Fig. 2 was developed directly from Fig. 1, for example, the 40,700 manhours at 30%

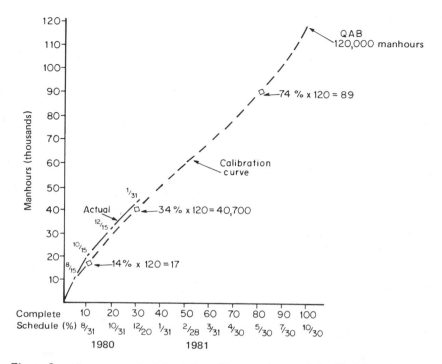

Figure 2   Above-ground piping labor. (Cost engineers, John Kashflow.)

complete were calculated by reading 34%, the theoretical percentage of budgeted manhours, from Fig. 1 and multiplying by the total QAB manhours, in this case 120,000. This type of plot gives us a calibration curve in manhours that is directly related to the theoretical productivity profile. When actual manhours are plotted on the same chart with the calibration curve, the PM can see at a glance how this activity compares to the QAB.

In Fig. 2 we see that 43,000 manhours were actually required to achieve 30% completion, which compares to 40,700 manhours on the calibration curve; in other words, the project is in trouble unless some corrective action can be taken to improve productivity. Also, from Fig. 2 we see that 30% complete was scheduled for December 20 but not achieved until January 31. By plotting schedule dates under percent complete and writing in the actual dates, one can show on the same chart that this activity is overrunning manhours and also behind schedule. Trend curves of this type, combining both schedule and work-hour information, are handy tools for presenting the cost/schedule picture at a glance.

FORECASTING FROM TREND CURVES

We can forecast from trend curves by using an approach similar to that used with productivity profiles: for example, let us suppose that a construction activity has a QAB of 120,000 workhours. What will be the total forecast if 43,000 manhours have been expended when the activity reaches 30% physical completion and the trend chart calibration curve is as shown in Fig. 2? Solve the problem as follows:

Actual manhours expended	43,000
Less theoretical manhours at 30% complete from the calibration curve	40,700
Overrun manhours	2,300
% Overrun = 2,300/40,700 × 100 =	5.7
QAB manhours	120,000
Overrun manhours = 5.7% of 120,000	6,800

Of course this answer assumes that no corrective action will be taken and that the actual manhours will maintain a constant percentage relationships with the calibration curve. Neither assumption will necessarily hold true. The first possibility is that corrective action will be taken by PMs when the potential overrun is called to their attention. The second possibility is that no corrective action will be possible and that the situation will deteriorate making the overrun larger. In any case this account should be carefully monitored by continuing to plot actual manhours and comparing to the calibration curve.

DANGER OF PLOTTING AGAINST TIME

In Fig. 2 we have plotted manhours against percent complete. Instead of percent complete we could have used time, as shown in Fig. 3. Here we have the same information as in Fig. 2. The original budget or calibration curve has been plotted against the same schedule as Fig. 2, but actual manhours are now plotted against time. Note that the shapes of the curves are now different and that the apparent overrun has been transformed into an underrun. This approach completely neglects the important aspect of schedule. Manhours are liquidated but the expected progress is not achieved. In an extreme case it would be possible to use all of the manhours in the QAB and make zero progress and the curve plotted against time only would give no indication of trouble. Some indirect activities are time-related and should be so plotted, but direct labor is progress-related and must be plotted against percent complete.

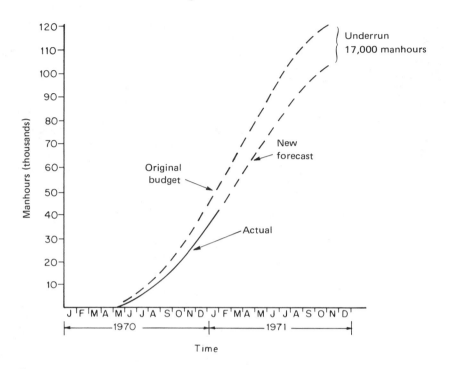

Figure 3   Above-ground piping labor. (Seat of the Pants, Inc., Engineers and Constructors.)

INTERRELATION OF MANHOURS, SCHEDULE, AND PERCENT COMPLETE

From the preceding discussion one can see the importance of the percent complete measurement. Theoretical (calibration) curves are not valid if the basis for percent complete changes. The method of labor forecasting presented herein is dependent on accurate and consistent progress measurement. The forecast interfaces with the schedule, and remember, a direct activity that overruns manhours and is behind schedule will probably cause a corresponding overrun in supporting indirect labor.

MORE ON LABOR PRODUCTIVITY

Figure 4 is a productivity profile used by the FCE on a recent major refinery project. The abscissa is, as usual, percent complete with the scheduled dates

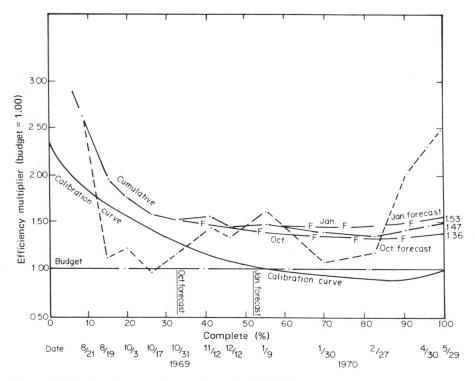

Figure 4   Productivity profile activity code 0216XX.

shown underneath, but the scale used for the ordinate is the reciprocal of productivity, frequently called the efficiency multiplier. When using the efficiency multiplier, actual or forecasted manhours always equals the QAB times the multiplier, and this concept is preferred by some companies. When reading productivity profiles using the multiplier, one must keep in mind that bad productivity is above the budget line and that good productivity falls below.

The dashed line on Fig. 4 represents period productivity, and it shows that budget productivity was reached only once during the project, at about 27% complete. After that productivity deteriorated until the 70% point where it once again approached the budget rate. Following the dashed line, one sees that the period between 46 and 55% was one of particularly bad productivity. At the 55% complete point, with an efficiency multiplier of 1.7, the PM took belated corrective action. He reorganized the craft involved, which had been working on an area basis with a small permanent gang in each construction area, into one gang that moved from area to area depending on the workload; also, he appointed a new craft superintendent and dismissed some less efficient workers. The result was the dramatic improvement in productivity that took place between 55 and 70% complete. The final productivity tail-off was as expected and it occurs in the last phases of nearly all activities.

The FCE's forecast lines in Fig. 4 are of particular interest. After productivity turned sour at 27% complete, the FCE made the October forecast, which predicted a multiplier of 1.36. At the peak of bad productivity (55% complete) he made the January forecast predicting a multiplier of 1.57. It was this forecast that triggered the PM's corrective action resulting in an improvement down to a final multiplier of 1.47. From the curve one can deduce that the PM's action saved six efficiency points; however, the question must be asked, if the PM had acted after the October forecast instead of waiting until January, would she not have saved a good deal more? From the chart it appears that the answer is yes; however, because of the work schedule it may not have been possible to take the same action in October that was effective in January. Another contributing factor could have been weather: was the October/January period one of bad weather and January/February a period of good weather? We can't make positive answers to these questions now, but we can be sure that the productivity profile triggered correction and that this action saved some manhours.

THE TREND CURVE

Instead of forecasting from the productivity profile as was done in Fig. 4, one could forecast equally well from the trend curve shown in Fig. 5. In the project from which these samples were extracted, the FCE actually drew both curves. The profile was used as a worksheet and kept in the FCE's notebook, whereas the trend chart was published monthly in the field cost report. In Fig. 5 we can see both the October and January forecasts. Changes in period productivity are reflected by the slope of the trend line: the flatter the slope, the better the period productivity. Also, from the dates shown on the trend line we see that the activity was on schedule through the 67% point but 15 days behind schedule at the 90% point. It appears that the action taken by the PM to reduce manhours caused a schedule delay, but in this case there was sufficient float so that delay in this activity has no impact on the project completion date.

The example illustrated in Figs. 4 and 5 is an extreme case. Normally we expect multipliers to be in the 5 to 20% range. Multipliers in the 40 to 50% range occasionally indicate the existence of an estimating or field timekeeping problem resulting in apparent bad productivity. Such was not the case in this example, but the FCE must always be alert for such possibilities.

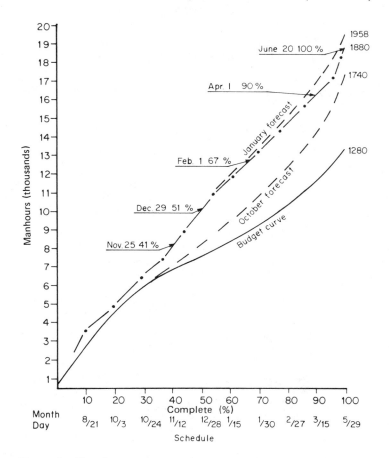

Figure 5 Trend curve.

SHAPE OF THE THEORETICAL (CALIBRATION) PROFILE

It must be emphasized that the shape of the theoretical profile is sensitive to the methods used for calculation of percent complete. If one uses data from a past project to establish the calibration profile for a current project, they must make sure that percent complete is calculated in the same manner on both projects. Also, one should not expect the profile for concrete to be the same as that for above-ground piping. Each major activity (i.e., concrete, structural steel, piping, etc.) has its distinctive calibration profile and the shape is sensitive to the method of calculating percent complete.

All activities can be combined into an overall profile for the entire project or for each major area. These overall profiles are good gross measuring tools providing percent complete is calculated on a consistent basis throughout. Examples of actual profiles from one project are shown in Fig. 6.

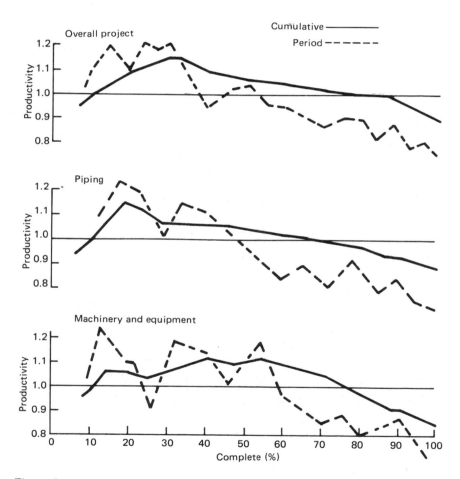

Figure 6   Examples of actual profiles from one project.

WORKHOURS FOR ONE PERCENT COMPLETION

In Chapter 15, we discussed the plotting of manhours for one percent completion curves (refer to Chap. 15, Fig. 1) and using them as forecasting tools. Some cost engineers like to use the same technique for direct field labor. As can be seen from Chap. 15, Fig. 1, one percent completion curves have the same general shape as productivity profiles plotted against the efficiency multiplier (refer to Fig. 4).

If there were no change orders and QAB manhours remained constant, the two curves would have identical shapes. The fact becomes clear after a quick look at the formulae:

1.   Productivity $= \dfrac{\% \text{ physical completion}}{\% \text{ of QAB manhours used}}$

2.   Productivity $= \dfrac{\% \text{ physical completion} \times \text{QAB manhours}}{\text{manhours used}}$

3.   Efficiency multiplier $= \dfrac{1}{\text{productivity}}$

4.   Efficiency multipler $= \dfrac{\text{manhours used}}{\% \text{ physical completion} \times \text{QAB manhours}}$

From Formula 4, one can see at a glance that the only difference between manhours per one percent complete and the efficiency multiplier is the divisor QAB manhours, and if the QAB manhours remain constant, the two curves will have identical shapes.

Obviously there is nothing wrong with using manhours per one percent complete as a forecasting tool. However, one must bear in mind that the QAB, the measuring stick, does not remain a constant but changes when change orders are added or when an activity planned for subcontract is shifted to direct-hire labor or vice versa. Since looking at manhours for one percent complete neglects changes (usually growth) in the yardstick (QAB manhours), most CEs prefer to work with productivity, and manhours for one percent complete remains a secondary tool.

PRODUCTIVITY EXPRESSED AS MANPOWER PER UNIT OF WORK

Previously we mentioned that productivity for a unit of work is equal to the budget unit rate divided by the actual unit rate. This is a straightforward way of measuring productivity and can be applied to activities with readily measurable quantities. Unit measurement can be used to compare the individual performance of work crews or to set targets for construction supervisors. The actual achieved unit rates can be plotted weekly and adverse trends quickly spotted. Figure 7 is an example of such a plot showing a comparison of the installation of fireproofing in three separate work areas each with its own supervisor. When properly used, charts like this can create competition and boost morale. They are good tools and should be used where applicable.

THE IMPACT OF OVERTIME

As mentioned earlier in this chapter, there is a decrease in labor productivity as the work week is extended. There is probably no productivity fall-off for occasional and emergency overtime, but when the scheduled workday is extended from 8 to 10 or 11 hr, workers tend to pace themselves. This tendency is magnified if the job is put on a scheduled 6- or 7-day week. Also, absenteeism increases significantly if workers are placed on a 6-day week over an extended period of time. The absenteeism further decreases productivity because crew sizes are disrupted and hurried makeshift reorganization becomes necessary. Many construction managers and superintendents fail to recognize the productivity debit and use overtime as a crutch to replace good planning and scheduling.

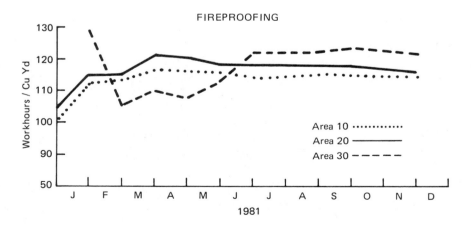

Figure 7   A comparison of the installation of fireproofing in three separate work areas each with its own supervisor.

Remember, you pay two premiums for scheduled overtime, first in increased wages and second in less productivity per manhour.

Although there is general agreement that prolonged overtime has a serious detrimental effect on productivity, there is not much hard evidence available to quantify this effect. There is some evidence that working a 60-hr week for a prolonged period of three months will reduce productivity by 35% compared to a normal 40-hr week and that even 50-hr weeks will reduce productivity by more than 25%.* As an example of the cost of overtime let us consider the following conditions:

1. After two months of steady 60-hr weeks, productivity has dropped by 15%, and by union agreement, double time is paid for overtime.
2. The percentage cost increase compared to a 40-hr week can be calculated as follows:

    a. 60 hrs are worked and productivity has dropped 15%.
        Equivalent hours of work = 0.85 × 60   =   51
    b. Equivalent hours gained = 51-40   =   11
    c. Additional hours paid   =   40
    d. % of straight time cost 40/11 × 100   =   360

3. If only time and one half were paid for overtime, the % of straight time cost would be 30/11 × 100   =   270

The preceding example illustrates the huge increase in cost due to overtime. Frequently a limited second shift will be cheaper and should always be considered before a job is placed on prolonged, scheduled overtime.

WORKER DENSITY

If skilled labor is available for the hiring, it is seldom economical to consider overtime before worker density begins to reach the saturation level. By saturation level we mean the point where the presence of other workers begins to seriously impact the normal productivity of each individual worker. There is always a great deal of noise and some interference on construction sites; however, when a number of crafts are working in the same area simultaneously and workers feel inhibited by people working above and close on each side, there is a productivity drop.

Density considerations on an overall basis should be a part of manpower planning, and the following guidelines have been used for refinery and chemical plant construction: If the working area including platforms is 300 SF or more per person, there is not likely to be density problems. If the area is between 200

*Business Roundtable Task Force Report, Nov. 1980

SF and 300 SF per person, there may be some productivity loss, and if the available area is less than 200 SF per person, a significant drop in productivity can be expected. When a project is falling behind schedule most construction managers will saturate the lagging areas with workers or put the project on scheduled overtime or do both. In such situations it is the FCE's job to study the conditions and advise the cost impact of scheduled overtime, second shift, area saturation, or accepting a schedule debit. Frequently it will be found that schedule debit is the most economical route from the standpoint of construction cost; however, other factors, such as value of the manufactured product, must also be considered before a final decision is made.

CREW SIZE AND LAYOFF SCHEDULE

Near the end of a large project or shortly past the midpoint on a small job, it becomes possible to forecast labor by crew size and labor rundown curves. When a project reaches the point where the remaining work can be visualized by the field supervisors, their estimates of crew days required for completion are usually more accurate than trend or productivity curves. This method is especially useful on small jobs where there has been no opportunity to collect the data required for use of the more sophisticated tools.

DIRECT-LABOR MANNING CURVES

It is probable that labor manning curves have been used since construction of the pyramids, yet even today they are frequently not used to full advantage. Figure 8 is an example of a labor manning curve as used for forecasting on a recent project. In Fig. 8 numbers of workers are plotted against time, and the original manning curve can be seen compared to a subsequent forecast and actual experience. Labor rundown was scheduled to start in mid-December, and when this did not occur, the PM was alerted to the probability of an overrun. As shown in the figure, a revised forecast, which turned out to be close to actual experience, was made in mid-February. Note that the project was completed 1.5 months behind schedule and that the second forecast, made in mid-February, recognized the schedule delay.

Manning curves for similar-type projects follow consistent patterns. Calibration curves can be determined from several past projects and, thereafter, forecasting curves can be laid out with fair accuracy knowing only the total estimated manhours and the schedule.

Figure 9 (page 246) is a sample calibration curve developed for one type of project. The key ratio of peak to average manpower is 1.54 in this example, and the labor centroid occurs at 54.8% of the construction time span. Average manpower as determined from the estimate and schedule equals 65 workers. By

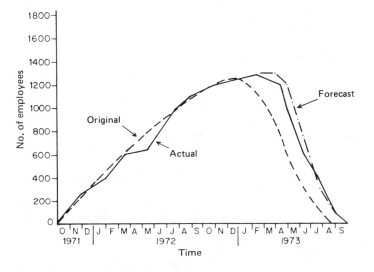

Figure 8   South Pole refinery labor manning curve.

using this idealized calibration curve developed from past projects, forecasting curves (such as Fig. 8) can be developed for future projects.

Normally, manning curves are drawn with calendar time as the abscissa and, of course, the number of days in the work week as well as the hours per day must be considered when laying out the curves. The peak-to-average ratio and distance to the centroid can vary depending on the type of project. Also, external restraints, such as availability of manpower, affect the curves. Actual manpower plotted on the same chart as the forecasted manning curve provides management with a simple but useful forecasting/scheduling tool, and it should be an integral part of cost control on every project, large or small.

MANNING vs. PRODUCTIVITY

Figure 10 shows the overall productivity profile, manning curve, and percent progress for a major refinery project. Although this data is from one project, it is typical of most projects—peak productivity occurs well ahead of peak manning, and at peak manpower the monthly rate of progress (as shown by the slope of the progress curve) is about the same as it was at lower manpower. In other words, increasing manpower from 4500 workers to 6000 workers appears to

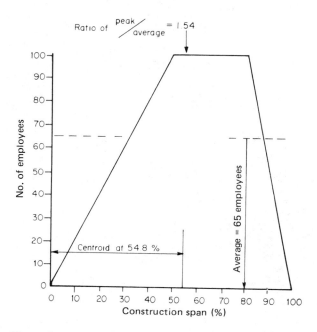

Figure 9   Idealized manning calibration curve.

have had little impact on the monthly completion rate. This anomaly is explained by the falling productivity. One must ask the question—would productivity have remained high and the monthly progress remained constant if the additional 1500 workers had not been hired? Since a project cannot be executed twice in two different ways, we do not have a positive answer to this question.

Every construction manager is faced with the question—how to maintain scheduled monthly progress with the leanest possible work-force? The theoretical answer is to maintain peak productivity for a long period of time; however, in practice there is no clear-cut right or wrong answer. The guidelines are the following:

Proceed with caution. Do not panic and overreact.

Increase the work force slowly and measure the impact on progress and productivity.

Take positive steps to increase productivity including awards or dinners for workers in areas showing high productivity, minimizing waiting time on tools and materials, providing the right tools, foreman training, and above all, efficient planning and scheduling.

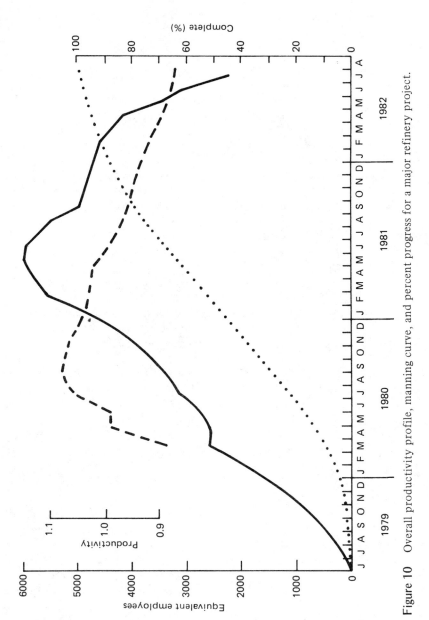

Figure 10   Overall productivity profile, manning curve, and percent progress for a major refinery project.

The CE's part is to provide the construction manager with the information needed. One way to do this is to show graphs similar to Fig. 10 every month. Figure 10 is for the overall project, but similar graphs can be prepared for individual activities such as concrete, structural steel, piping, or for individual construction areas.

MECHANICAL COMPLETION AND BUT LISTS

As we have seen, productivity usually decreases dramatically for all activities after the 90% completion point. This is largely due to correction of the myriad of minor oversights and errors that are discovered during final plant checkout prior to mechanical completion and acceptance by the owner. Craftsmen, who are skilled at wrestling a 20-in. dia. pipe spool into place at 100 ft. above the ground, are prone to overlook the importance of a ½ in. nipple and valve required for an instrument connection or vent line. Also, some things always fail during hydrotest and motor run in—a valve leaks here, a faulty weld is discovered there, and the large compressor turbine has excessive vibration. All of these things must be fixed before the plant can be declared complete.

As construction work nears completion, the startup team moves onto the site. These people have the responsibility of confirming that everything called for on the drawings or in the job specifications is not only in place but is also in working order. When the construction superintendent thinks he has an area or major piece of equipment mechanically complete, the start-up crew makes a complete check. Any errors, oversights, or problems are put on a punchout or but list signifying—"yes, you are mechanically complete *but* for these items." The superintendent then moves his people back into the area and works off the but list. The checkout is repeated until everything is found complete and in working order. Initial startup is then undertaken and the superintendent normally must keep a special crew available to assist if any deficiencies unexpectedly crop up.

But lists usually include many small items which are time consuming and tedious to correct. Frequently scaffolding must be reinstalled and working around existing equipment is slow and inefficient. Experience has shown that this work is usually best handled by special start-up crews including first class pipefitters, welders, electricians, and instrument technicians.

Some progress measurement systems attempted to compensate by withholding a small completion percentage to cover but list work. In other words, when a superintendent reports something is complete, he is given only, say, 98% progress with 2% held back for but list work. However it is done, it is important that the progress measurement be consistent. The manpower forecast must include an allowance for but list work. One reasonable way to do this is to assume that the productivity tailoff will be similar to past projects as reflected in the productivity calibration curve.

AVERAGE WAGE

Forecasting manhours is only the first part of predicting labor costs; the second part is forecasting wages. A fine manhour forecast is not worth much if it is multiplied by the wrong average wage rate. As is the case with productivity, the average wage rate varies over the life of a project. The following factors influence wage rates:

Amount of overtime hours: Since overtime is paid at time-and-a-half or double-time it has a tremendous impact.

Merit increases and promotions: There are usually several categories of laborers and apprentices, and as the job progresses the best of these people are promoted up the scale.

Labor contracts and negotiations: Frequently, during the life of a project, it is necessary to renegotiate wage agreements with labor unions. This always results in an increase.

Other projects in the same general area: Several simultaneous major projects within the same commuting area tend to drive wages upward.

Craft-to-helper ratio and skilled-to-unskilled-labor mix: As a project progresses, the labor mix changes from the less skilled to the more skilled trades; and in open-shop areas, out of union control, the journeyman-to-helper ratio has a big impact on the average wage.

The forecasting of wages requires that FCEs consider all known factors influencing wages in their area. There are few guidelines that can be given. The forecaster must rely on good judgement and common sense. If a general wage increase is scheduled during the project, any event that moves the labor centroid (refer to Fig. 9) will have an impact on average wage. If in the early phases actual manning lags the forecasted curve, the effect will be to move the labor centroid to the right, and work not accomplished in the early phases at a low wage rate will be accomplished later at a higher rate. The manning curve is a useful tool in forecasting wage rate.

Figure 11 illustrates how wages increased over the life of a recent project. In the example shown, the final cumulative wage rate was 20% higher than the average rate paid at the start of the project.

SUMMARY

Forecasting direct-labor manhours is probably the most difficult job facing the CE. As we have seen in this chapter the available forecasting tools include the following:

The original budget estimate: this is the first forecast.

The QAB: this is a recast budget using original budget productivity ratios multiplied by actual designed quantities.

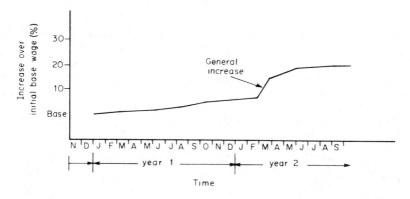

Figure 11   Average construction wages, South Pole refinery. (Prepared by John Dollarspotter.)

Productivity profiles: budget productivity rates are modified to reflect actual project experience.

Manning curves: forcasted manpower and schedule are compared to actuals.

Crew sizes and layoff schedules: these become important during the final stages of construction.

But list work must be reflected in both the manpower forecast and construction schedule.

The second part of the labor forecast—wage rates—requires sound judgment and a knowledge of craft wage rates and labor mix. Wage rate tables are frequently negotiated with the union at the start of a project; in any case, craft rates are always available from the payroll. The anticipated labor mix can be obtained from the control estimate, and thereafter it must be monitored to reflect actual project experience. Overtime is costly, use it wisely and always consider the alternatives.

Direct labor is not the only problem facing the FCE, and other important cost elements will be discussed in the next chapter.

22

COST CONTROL DURING CONSTRUCTION: ITEMS OTHER THAN DIRECT FIELD LABOR

INTRODUCTION

As mentioned in Chap. 20, cost elements other than direct labor that require control include

Bulk materials and surpluses
Field purchases
Subcontracts
FLOH
Miscellaneous overheads

In this chapter we will discuss methods used by the FCE in providing management with the information required to exercise cost control over these elements of expenditure. Many construction management organizations focus on direct labor and neglect the large dollar amounts that are expended on these other essential but "extracurricular" activities. It is the FCE's job to point out the importance of these activities and to aid in controlling them. The first item requiring attention is bulk materials.

BULK MATERIALS

Bulk materials purchased from the engineering office are stored in the field warehouse until required for construction. The ideal situation is to schedule deliveries so that most materials can bypass the warehouse and be unloaded directly at the construction site as they are needed; unfortunately, this ideal is seldom realized in practice and the majority of bulk materials wind up on the warehouse yard where they must be maintained free of rust or other damage and accessible when required. Entire books have been written on the storage and handling of materials, and material-handling labor is a significant indirect account that requires careful control.

The FCE should investigate the material-handling procedures and verify that accurate records are kept of materials delivered, items on hand, and items issued to the site for erection. If good records are not kept pilferage may occur on a large scale and unexpected shortages, necessitating costly emergency purchases or delays, will occur.

Many items, including reinforcing bar, form lumber, tie wire, anchor bolts, shim stock, and electrical wire are purchased directly by the field. Field superintendents have a tendency to purchase everything on an emergency basis, or else, fearing to run short, they purchase a huge stock that will be left surplus at the end of the job. Good management requires that field requirements be forecasted in advance and that time be allowed for obtaining competitive bids. Field superintendents should be required to submit justification when they overrun their forecasts; also, consumption rates must be watched to prevent accumulation of surpluses. On lump sum or hard-money jobs contractors will control their field purchases, but in reimbursable situations it's indeed a rare contractor that controls field purchases without considerable prompting by the owner. It is the FCE's responsibility to verify that field purchases are tightly controlled, to monitor costs, and to advise of forecasted over- or underruns. This is usually done by the use of a trend curve for each field purchase account.

SURPLUS MATERIALS

No matter how carefully bulk materials are controlled, appreciable quantities always wind up as surplus. Surpluses come about in the following ways:

Late changes to the project.

Overpurchasing to make sure that there will be no shortage affecting the construction schedule.

Material takeoff and purchasing errors (i.e., 1000 widgets purchased when only 10 are required).

Early purchases made by management before engineering takeoffs are available. This is occasionally done when a period of high inflation and material shortages is anticipated for the immediate future.

Small-bore pipe, fittings, and valves usually make up 50 to 75% of the total surplus value. Other items include:

Reinforcing bar
Form lumber
Electrical wire and conduit
Insulation

As discussed in Chap. 17, final bulk-material takeoffs and commodity reviews are usually made in the home office when engineering is about 95% complete; however, most bulks will have been ordered long before that time with quantities based on preliminary takeoffs. The final takeoffs are intended to spot shortages and surpluses so that bulk orders can be finalized. Standard-type surplus items can frequently be returned to the supplier with the payment of a restocking charge. However, this information usually becomes available when construction activity is at its peak and nobody has time to worry about surplus materials. Therefore, it is frequently left up to the FCE to prod purchasing and material-control personnel into action and see that surpluses are returned and credit received.

Most contractors have sophisticated systems for controlling bulks that operate as follows using rebar as an example:

From preliminary takeoffs initial purchases are made and purchased quantities and sizes are entered into a computer.

As detailed design progresses, exact quantities for each foundation or concrete structure are also entered into the computer.

At any time the computer can then print out quantities ordered, quantities required, and remaining surplus quantities.

This is a good system, enabling engineers to minimize surpluses by tailoring their final designs to use the remaining sizes of rebar. Similar computerized systems are used for piping, electrical, and other bulk materials; however, for some reason known only to computer gods and not divulged to mere humans, there is always a discrepancy between what is actually on hand in the field and what is on paper in the office. For this reason when construction is about 75% complete and everyone is worrying about construction schedule, plant startup, or their next assignment FCEs should

Insist that bulk inventory be taken and results compared to remaining requirements

Initiate action for disposal of obvious surplus

Include a realistic value for sale of surplus in their cost forecast

Initiate action for disposal of purchased construction equipment and unrequired
hand tools

FIELD LABOR OVERHEADS

FLOH ranks second only to direct labor on the FCE's priority list. On many proj-
ects *the cost of FLOH is equal to direct labor,* but it is seldom subjected to the same analysis and control. As we have seen, these indirect costs fall into three distinct groups as shown in Table 1. Many overhead categories are time-related

Table 1   FLOH Categories

Temporary construction and consumables

    Temporary buildings
    Temporary utility lines
    Utilities used during construction
    Material handling and storage
    Consumable supplies and safety equipment

Construction tools

    Equipment (cranes, dozers, gin poles, etc.)
    Small tools
    Equipment maintenance and fuel
    Trucks and autos

Supervision

    Craft and area supervisors
    Field-office staff
    Surveyors
    Guards and janitors

---

and the general approach is to monitor them by trend charts with time as the abscissa and dollar cost as the ordinate. In the next few paragraphs we will consider each of the major categories separately.

TEMPORARY CONSTRUCTION

On a reimbursable project the owner should insist that the contractor supply sketches, layouts, and cost estimates for temporary facilities before moving onto the site. The following items should be included:

Temporary office, carpenter shop, machine shop, pipe shop, warehouse, automotive and equipment repair shop, gatehouse, and field toilets
Temporary roads, material storage areas, fencing, and parking areas
Concrete batch plant or other source of concrete
Temporary power lines, transformers, water, sewer, and air lines

The contractor's package should include a plot plan showing a dimensional layout of the preceding facilites. The estimates should be in considerable detail presenting quotes or quantitative detail for the office, warehouse, and other buildings. Items such as office furniture, typewriters, calculators, and warehouse shelving must not be overlooked. Also, the estimate must allow for maintenance and dismantling. After careful review, the contractor's estimates can be accepted as the control budget. Thereafter, this budget is used as the base for forecasting overruns or underruns. If contractors later propose items not in the budget, they should submit complete justification to the owner.

On lump sum projects contractors will tightly control the temporary facilities and their CEs or superintendents will be held strictly accountable.

From a control standpoint, temporary facilities should be looked at from both quantity and cost viewpoints. Office area in the budget should be compared to the area actually built and overruns in area must be justified even though the total cost of the buildings is an underrun.

Most contractors have standard cost codes adequate for the control of temporary facilities. The FCE must review these codes, verify that they are adequate, and make sure that there is a budget for each applicable code.

CONSUMABLE SUPPLIES AND MATERIAL HANDLING

Consumable construction supplies are a function of the size of the labor force and are usually estimated as a percentage of direct-labor cost. The percentage, derived from past projects, can be deceiving if labor wages and material prices are escalating at different rates. In such situations FCEs must make sure that their budget is adjusted to reflect material and not labor escalation. Material-storage and handling costs are also estimated as functions of the direct-labor cost and in this case there is not differential escalation problem since material handling is a labor operation. These two items are best controlled by trend curves with dollar expenditure for consumables and manhours for material handling plotted against time. The expenditure pattern (calibration curve) determined from previous projects is plotted first, and as the project progresses actual expenditures are compared to the pattern and over- or underruns forecasted. Figure 1 is a trend curve showing forecasted and actual expenditures for consumables on a recent project.

A curve such as Fig. 1 is used to highlight problem areas. Whenever actual expenditures exceed the pattern (calibration curve), investigation is called for. The investigation has three purposes:

1. To determine the cause of the excessive expenditure
2. To take corrective action to minimize future expenditures
3. To reforecast the total cost

Typical reasons frequently causing overruns of consumables include the following:

1. Building up a large warehouse inventory
2. Lack of control of quantities issued such as the following:
    a. One worker issued five pairs of boots
    b. Excessive issue of work gloves
    c. New rain-protective gear issued every time it rains

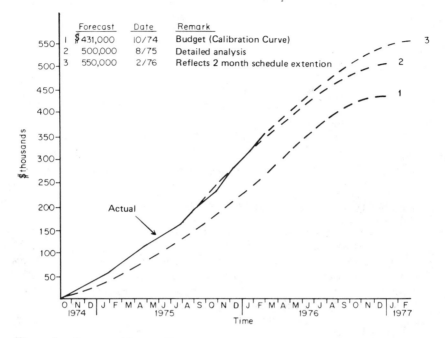

Figure 1 Consumable supplies.

    d.   Poor purchasing practices and failure to obtain competitive bids

    e.   Pilferage

In the case of consumables some corrective action is nearly always possible and the same applies to material handling. Good management in these areas can save hard dollars for the owner or contractor.

UTILITIES USED DURING CONSTRUCTION

Normally the largest utility cost is electricity, but other important items include water for hydraulic testing of piping and vessels, water for concrete, and drinking water. In some cases disposal of sanitary sewage becomes a costly problem. The utility accounts are best monitored by trend curves of the type just described for consumable supplies.

CONSTRUCTION TOOLS

As shown in Table 1, construction tools comprise four subaccounts; of these small tools and equipment maintenance are best monitored by trend curves. Normally there is a resale or recovery value on small tools, which should be forecasted on the trend curve. Equipment maintenance can be divided into three detail accounts as follows:

1. Spare parts
2. Fuel, oil, and grease
3. Service and repair labor

Trend curves are not adequate for control of large equipment pieces, trucks, and autos. Generally, this equipment is rented; however, on large projects, purchase may be economical. On a reimbursable project, before making a decision to buy or rent, the owner should consider purchase price, rental rate, estimated time on the job, time value of money, and probable resale value. The best rental rates are usually obtained by soliciting competitive bids for supplying all the construction equipment. Often, the construction contractor will have equipment available and offer it at fixed rates. When project costs are fully reimbursable the owner should ask contractors to submit their bids along with that of other suppliers. Frequently competitive bids will save the owner 5 to 15% on equipment rental.

For control of construction, shop, and automotive equipment an equipment schedule should be prepared showing each major equipment category as a separate line. Figure 2 is one page from the equipment schedule used on a recent project. Considering item 2, 200-ton crawler crane, we see that the budget included one purchased unit and five rental units for a total rental period of 49 crane months. From the forecast we see that one unit was indeed purchased and a maximum of five planned for rental; however, the crane rental months have been reappraised and are forecasted to total 47 or 2 crane-months under the budget. The underrun in months is further improved by a slight underrun in rental rate, and the total forecast for 200-ton cranes, including purchased and rental, is $469,550 vs. a budget of $485,100.

With the data available in Fig. 2, cost control of construction equipment becomes an easy task. The control chart should be set up by the construction manager and the FCE before moving to the field. Thereafter, it should be updated frequently and the forecast revised to reflect actual project execution plans as they are developed.

FIELD SUPERVISON

The number of persons on the contractor's field supervisory staff must be carefully watched from the cost standpoint. Before moving into the field, contractors

No.	Equipment Piece			1975 J	F	M	A	M	J	J	A	S	O	N	D	1976 J	F	M	A	M	J	J	A	S	O	N	D	Rental Mo.	$ Rate	Total	Net Purch Cost[a]	Dollar Total Cost
1	Crane	Bud.	Buy																													
	Crawler		Rent	2	2	2	2	2	2	2	2	2	2	2	2	2	2	2	2	2	2							36	18500	666000		666000
	350 Ton	F'cast	Buy																													
	Manitowoc		Rent	1	2	2	2	2	2	2	2	2	2	2	2	2	1	1	0	1	2	1	1	1	1			34	18500	629000		629000
2	Crane	Bud.	Buy	1	1	1	1	1	1	1	1	1	1	1	1	1	1	1	1	1	1	1	1	1	1	1					93100	
	Crawler		Rent			1	2	5	5	5	5	5	5	5	3	3	3	1	1									49	8000	392000		485100
	200 Ton	F'cast	Buy	1	1	1	1	1	1	1	1	1	1	1	1	1	1	1	1	1	1	1	1	1	1	1					95900	
	Manitowoc		Rent				1	4	5	5	5	5	5	5	3	3	3	2	1									47	7950	373650		469550
3	Crane	Bud.	Buy																													
	Crawler		Rent				1	4	4	4	4	4	4	4	4	4	4	3	2	1	1	1	1	1	1			52	6400	332800		332800
	150 Ton	F'cast	Buy																													
	Manitowoc		Rent						3	4	4	4	4	5	5	5	4	4	4	3	2	2	2	1	1	1		58	6000	348000		348000
4	Crane	Bud.	Buy	1	1	1	1	1	1	1	1	1	1	1	1	1	1	1	1	1	1	1	1	1	1	1					40200	
	Crawler		Rent		1	5	5	6	6	8	8	8	8	8	6	6	3	3	3	2	2	1						93	3600	334800		375000
	70 Ton	F'cast	Buy		1	1	1	1	1	1	1	1	1	1	1	1	1	1	1	1	1	1	1	1	1	1	1				51300	
	American		Rent		1	2	2	3	4	7	7	8	8	8	8	6	5	5	3	3	2	1	1					84	3300	277200		328500
5	Crane	Bud.	Buy	1	1	1	1	1	1	1	1	1	1	1	1	1	1	1	1	1	1	1	1	1	1	1	1				39100	
	Motor		Rent							1	1	2	2	2	2	2	2	2	2	2	2	2	2	1	1			28	1800	50400		895000
	45 Ton	F'cast	Buy		1	1	1	1	1	1	1	1	1	1	1	1	1	1	1	1	1	1	1	1	1	1					42500	
	American		Rent								1	1	2	2	2	2	2	3	3	3	2	2	1	1	1			28	1700	47600		90100
6	Crane	Bud.	Buy	3	6	8	10	10	12	12	12	12	12	12	12	12	12	12	12	12	12	12	12	8	8	4	4				144000	
	Hydraulic		Rent							1	4	8	11	11	15	15	15	10	10	6	4	2	1					113	1450	163850		307850
	15 Ton	F'cast	Buy	3	6	6	6	10	12	12	12	12	12	12	12	12	12	12	12	12	12	12	12	12	8	6	6				156000	
	Cherry Picker		Rent							2	4	10	12	12	16	16	16	16	12	10	6	4	2					138	1400	193200		349200
7	Reactor	Bud.	Buy																													
	Transporters		Rent									1	1	1	1	1												5	15000	75000		75000
	Motorized	F'cast	Buy																													
	(pair)		Rent										1	1	1	1	1	1			1	1						8	16500	132000		132000
																												Page total budget				2331250
																												Page total forecast				2346350

[a]Purchase cost less estimated resale.

Figure 2 One page from the equipment schedule used on a recent project.

should prepare an organization chart showing each member of their staff and the estimated time on the job. This information can be put into a bar chart forecast similar to that used for construction equipment. The bar chart, called the personnel schedule, should include the various categories of people and the estimated workmonths in each category. As personnel move onto the job, the FCE compares the actual persons in each category with the estimate. Potential overruns can be quickly spotted on such a bar chart. When a potential overrun occurs, the workload of the individuals in that category should be examined to see if a possible savings can be made. Average salaries for each personnel category can be added to the bar chart, thus making it easy to update the forecast to reflect changes in personnel. When forecasting salaries, the FCE must remember to include escalation. It is only at the exact project centroid that current salaries paid will equal the average for the project. Figure 3 is a sample personnel schedule as used on a recent project.

The personnel schedule includes not only the construction supervisors but also support personnel, such as timekeepers, clerks, stenographers, office engineers, FCEs, surveyors, and inspectors. From the personnel schedule the FCE can prepare a trend chart for the cost of all field staff. The budget curve calculated from the personnel schedule can be directly compared to actual expenditures on the chart and errors or omissions on the personnel schedule will be highlighted.

OTHER OVERHEADS

Every project has a number of miscellaneous overheads peculiar to that project and its location. These overheads may include the following:

Guards
Janitors
Ice-water distribution
Freight
Staff travel
Cost of selling surplus
Recruiting
Computer costs
Payroll handling
Rework
Back charges

Since every project is different, it is impossible to make a complete list of miscellaneous overheads, but the contractor's standard cost code is usually a good checklist and it should be carefully reviewed to make sure that all applicable accounts have a budget.

Figure 3 — A sample personnel schedule as used on a recent project.

No.	Description		1975 J	F	M	A	M	J	J	A	S	O	N	D	1976 J	F	M	A	M	J	J	A	S	O	N	D	Work Months	Monthly Rate	Total Cost (Dollars)
1	Construction Manager	Bud.	1	1	1	1	1	1	1	1	1	1	1	1	1	1	1	1	1	1	1	1	1	1	1	1	24	2600	62400
		F'cast	0.5	0.5	1	1	1	1	1	1	1	1	1	1	1	1	1	1	1	1	1	1	1	1	0.5	1	21.5	2800	60200
2	Assistant Construction Manager	Bud.			1	1	1	1	1	1	1	1	1	1	1	1	1	1	1	1	1	1	1	1			20	2500	50000
		F'cast	1	1	1	1	1	1	1	1	1	1	1	1	1	1	1	1	1	1	1	1	1	1	1	1	23	2500	57500
3	General Superintendents	Bud.					2	2	2	2	2	2	2	2	2	2	2	2	2	2	2	2	2	2	2	2	39	2100	81900
		F'cast			1	2	2	3	3	3	3	3	3	3	3	3	3	3	3	2	2	2	2	2	1	1	48	2000	96000
4	Rigging Superintendent	Bud.	1	1	1	1	1	1	1	1	1	1	1	1	1	1	1	1	1	1							18	1800	32400
		F'cast	1	1	1	1	1	1	1	1	1	1	1	1	1	1	1	1	1	1	1						19	1850	35150
5	Assistant Rigging Superintendents	Bud.	1	1	2	2	2	2	2	2	2	2	2	2	2	2	1	1									31	1700	52700
		F'cast	1	1	1	1	2	2	2	2	2	2	2	2	2	2	1	1									32	1650	52800
6	Boilermaker Superintendent	Bud.		1	1	1	1	1	1	1	1	1	1	1	1	1	1	1	1	1	1	1					20	1800	36000
		F'cast			1	1	1	1	1	1	1	1	1	1	1	1	1	1	1	1	1	1	1				20	1850	37000
7	Assistant Boilermaker Superintendents	Bud.						1	1	2	2	2	2	2	2	2	2	2	2	2	2	1					30	1700	51000
		F'cast						1	1	2	2	2	2	2	2	2	2	2	2	2	2	2	1				30	1650	49500
8	Electrical Superintendent	Bud.	1	1	1	1	1	1	1	1	1	1	1	1	1	1	1	1	1	1							18	1800	32400
		F'cast	1	1	1	1	1	1	1	1	1	1	1	1	1	1	1	1	1	1	1						19	1900	36100
9	Instrument Superintendent	Bud.					1	1	1	1	1	1	1	1	1	1	1	1	1	1	1						17	1800	30600
		F'cast					1	1	1	1	1	1	1	1	1	1	1	1	1	1	1						17	1900	32300
10	Insulation and paint Superintendent	Bud.								1	1	1	1	1	1	1	1	1	1	1	1	1	1				15	1700	25500
		F'cast							1	1	1	1	1	1	1	1	1	1	1	1	1	1	1				16	1700	27200
11	Piping Superintendent	Bud.			1	1	1	1	1	1	1	1	1	1	1	1	1	1	1	1	1	1	1	1			21	1900	39900
		F'cast		1	1	1	1	1	1	1	1	1	1	1	1	1	1	1	1	1	1	1	1	1			22	1850	40700
12	Assistant Piping Superintendents	Bud.				1	1	1	2	2	2	3	3	3	3	3	3	3	3	3	3	2	2	1	1		47	1700	79900
		F'cast				1	1	1	1	2	2	3	3	3	3	3	3	3	3	3	3	3	3	2	2		46	1650	75900
13	Chief Field Engineer	Bud.	1	1	1	1	1	1	1	1	1	1	1	1	1	1	1	1	1	1	1	1	1	1	1	1	24	2000	48000
		F'cast	1	1	1	1	1	1	1	1	1	1	1	1	1	1	1	1	1	1	1	1	1	1	1	1	24	2000	48000
14	Field Engineers	Bud.	2	2	3	3	3	4	4	4	4	4	4	4	4	4	4	4	4	4	4	3	3	3	2	2	81	1700	137700
		F'cast	1	1	2	3	3	3	4	4	4	4	4	4	4	4	4	4	4	4	4	3	3	3	2	2	77	1700	130900

MORE ON TREND CURVES

Depending on the nature of the job and the breakdown in the contractor's cost code, there may be from 50 to 100 field labor overhead (FLOH) accounts, and each of these accounts should be monitored by an individual tracking curve. Some accounts, such as vehicle maintenance, have both labor and material components. In these cases one tracking curve reflecting manhours should be set up for the labor portion, and another curve reflecting dollars used to track the material portion. In a few cases several small accounts administered by the same supervisor might be combined into one curve, but experience has shown that generally it is easier and more useful to plot each account separately.

The initial preparation of a budget calibration curve for each FLOH account is a great deal of work for the FCE. It should be done in the home office as soon as the detailed control estimate has been completed. However, after the curves have been set up, the plotting of actual costs in the field is easy. The curves used to track FLOH in the field are analogous to those used in the home office to track engineering expenses (refer to Chap. 15). The techniques for establishing calibration curves and comparing with actual expenditures is exactly the same as described in Chap. 15.

Figure 4 is an example of a calibration curve showing the rate of expenditure for one FLOH account for field office costs. The envelope has been developed by plotting the rates of expenditure from several previous projects. If actual costs of a current project fall within the envelope, no action need be taken, but if they fall outside the envelope, investigation is required to determine the reason. Some CEs prefer to use the calibration envelope and others prefer a single line using the centroid of the envelope as an average anticipated rate of expenditure. Many contractors develop standard control tools similar to Fig. 4 for each FLOH account. Note that percent of total cost has been plotted against percent of total construction time. To recast this curve to fit any specific project, one must know only the total estimated cost and the scheduled construction time span.

CONTROL OF FIELD LABOR OVERHEADS

Trend charts and bar charts are reporting and monitoring tools but they do not control. As mentioned previously, the cost of FLOH is in the same magnitude as direct labor and requires the same amount of management attention. Some overhead accounts, such as warehousing and road maintenance, are clearly the responsibility of individual supervisors; however, other accounts, like job cleanup and office supplies, do not have individual supervisors and receive charges from many different departments. In these situations, control and accountability are frequently absent and costs get out of hand.

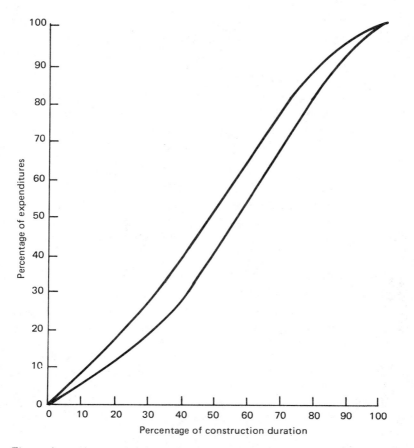

Figure 4   An example of a calibration curve showing the rate of expenditure for one FLOH account for field office costs.

Experience has shown that any construction account, with no specific person responsible for costs, will tend to overrun. When pressure to improve direct labor productivity is put on superintendents, they soon discover that more manhours charged to overhead accounts mean less hours charged to direct labor and an apparent improvement in productivity. Of course, such action is not condoned by upper management and is always vigorously denied by all concerned, but it happens—sometimes to an alarming degree.

One method of control, which has been found successful in the past, is to assign a high ranking management person to be responsible for each account. When the assigned person has direct supervision and control over the account, this method works very well. For other accounts, which receive charges from several departments or sections, the responsible person must be the construction manager or the assistant manager. The manager then assigns quotas to the departments charging the account. The CE monitors the quotas and provides feedback to the manager.

On reimbursable projects it is usually necessary for the owner to provide the impetus required for control of the overhead accounts. On a small project this may be handled by the owner's CE, but on a large project owners will find it worth their while to assign a high ranking individual full-time to cost control of FLOH. These persons, working closely with the owner's CE and the construction management staff, can save many times their salary. Frequent meetings between the owner and contractor to review the indirect tracking curves are essential. It goes without saying that the success of this kind of program requires the full and vigorous support of the owner's and contractor's project managers.

REWORK

Under the program for forecasting and control of direct labor outlined in Chap. 21, a rework account becomes a necessary evil. Rework can be defined as time spent to correct somebody else's error. It can be divided into two distinct categories or subaccounts—construction error and engineering error. The meaning of rework and the two categories are probably best defined by typical examples as follows:

1. The survey crew made an error in staking out the location of a pump foundation. Unaware of the error, the superintendent of concrete proceeded to form and pour the foundation and took full progress credit for the completed work. Subsequently the error was discovered and it became necessary to destroy the recently poured foundation and to pour it again in the correct location. This is an example of construction error. It is not correct to penalize the productivity of the concrete crew for work required to correct the surveyor's error; therefore, the workhours required for breaking out the erroneous foundation and repouring it should be charged to rework.

2. A piping designer made a dimensional error on a piping isometric drawing. Pipe spools were shop fabricated according to the drawing. The error was discovered in the field and it was necessary to modify the spools. This is an example of engineering error and the manhours required to correct the piping should be charged to rework.

3.   A welder made several faulty welds while connecting piping in the field. The faults were discovered during hydrotest. Manhours required to correct the welds should *not* be charged to rework. This is an example of extra work required by a craft to correct its mistakes.

Top management of some engineering/construction contractors are reluctant to set up rework accounts. They prefer to bury their mistakes and take their lumps as bad productivity which can always be blamed on the work force or union. This makes labor forecasting and control difficult by obscuring productivity problems; also, it provides craft superintendents with ready excuses for poor productivity. The lack of rework accounts may be an indication of sloppy management with little emphasis on productivity measurement and improvement.

Rework accounts are similar to FLOH accounts and must be carefully monitored to prevent abuse. Rework can be handled on a work order system whereby the involved superintendent submits a form describing and justifying the rework along with an estimate of the required workhours. After approval by the construction manager, the CE assigns a rework number and the superintendent is advised to proceed.

BACKCHARGES

Backcharges are claims against vendors for the supply of defective or off-specification equipment. Backcharges arise from defects discovered in the field during construction and are not to be confused with manufacturer warranties which normally apply after equipment has been placed in operation. Most backcharges arise from vendor errors in the fabrication of piping, structural steel and vessels. Common fabrication defects are flanges welded out of plumb with the pipe centerline, bolt holes in structural steel drilled in the wrong location, brackets omitted from vessels, and out of level supports in towers. Also, defects are sometimes found in specialty manufactured equipment including items such as wrong type of seal in pumps, wrong alloy used for impellers, and valves with parts made from the wrong alloy.

Although shop inspectors and quality control people do their best, some defects always slip through and must be corrected in the field; therefore, it is necessary to set up strict backcharge procedures before fabricated materials begin to arrive at the job site. The procedures should include the following:

Appointment of a backcharge coordinator. This individual is responsible for collecting and organizing all backcharge information. On a large project this will develop into a full-time job.

Timely advising of vendors as soon as defects are discovered. This is most important if backcharges are to be collected. Some vendors will send representatives to verify the defect and supervise repairs. Others will make the repairs with their own personnel, and still others will ask for an estimate and authorize the construction contractor to make repairs.

Setting up a complete file on each backcharge including a copy of the first notice to the vendor, photographs of the defect, estimated repair cost, copies of time sheets of personnel making the repair, copies of material requisitions, copies of all correspondence with the vendor, and calculation of the total cost including a markup for overhead.

Maintaining a backcharge log showing the status and value of each backcharge.

Backcharges are important to FCEs and they must work closely with the backcharge coordinator. In some organizations the backcharge coordinator reports to the lead CE. The CE's monthly cost report should show the total value of backcharges to date and the forecasted recovery. Until experience begins to indicate otherwise, it is reasonable to forecast no more than a 50% recovery overall. Also, the manpower forecast must include an allowance for craftsmen assigned to backcharge work. This work should be treated like an overhead item, and workers engaged on backcharges should not be reflected in direct labor productivity.

Backcharges are frequently settled by the simple expedient of reducing the amount to be paid to the vendor (i.e., reduce the purchase order value). The CE must work with accounting and make sure that this is properly cost coded and that backcharges are not used to reduce the value of the equipment purchased. A backcharge has nothing to do with the value of a piece of equipment; therefore, a special account should be set up to accumulate the monies charged to and collected from backcharges. Backcharges should appear as a separate item in the monthly cost forecast and not shown as a credit against any other cost element.

FIELD SUBCONTRACTS

In the United States many large contractors engaging in refinery and chemical-plant construction directly hire the great majority of their labor; however, local specialty subcontractors are always employed to some extent. In Europe the reverse is true; major contractors subcontract the majority of their labor and directly hire few workers. Frequently construction contractors fail to consider the cost advantages that can be gained by using local specialty subcontractors. On a reimbursable project, the owner's CE is in an ideal spot to insist that the economics of subcontracts be considered for special situations. The approach is to request general contractors to independently estimate the cost of performing a special function with their forces and compare this cost to bids solicited from available subcontractors. Special functions frequently subcontracted include the following:

Field-stress-relieving of piping
X-raying of welds

Insulation
Painting
Application of refractory
Erection of buildings
Site grading
Control-house wiring
Concrete testing
Fencing
Guards

The list can be long and varied and depends on the expertise of the general contractor and the skills of available subcontractors. In Europe, as indicated above, virtually all field activities are subcontracted. Subcontract development and control are discussed in detail in Chap. 19.

SUMMARY

The trend chart, often called the tracking curve, is the primary tool for cost control of overhead accounts. A tracking curve should be prepared for each major overhead account. The curves are usually time-related, and expenditure patterns (calibration curves) can be developed from past projects. Each tracking curve should show the budget and anticipated expenditure pattern. Actual expenditures are plotted, compared to the anticipated pattern, and budget over- or underruns forecasted. Old forecasts are not erased but left on the chart as shown on Fig. 1. Manhours rather than dollars should be the ordinate of trend curves for indirect-labor accounts, such as material handling. Average wages can be plotted on the same chart, but to a separate scale, thus graphically illustrating the status of the entire account on one piece of paper.

Special tools for large accounts include the construction equipment schedule (Fig. 2), the personnel schedule (Fig. 3), a detailed dimensioned layout of temporary facilities, and the subcontract register.

The FCE works closely with the subcontract administrator and makes sure that the subcontract procedures include the elements necessary for cost and schedule control. The maintenance of an accurate, up to date subcontract register is an important tool for the FCE, and close monitoring of subcontract schedules is indispensable.

A rework account is a necessary evil. It is needed to measure worker productivity and to maintain a degree of superintendent accountability for productivity of workers.

Backcharges tend to be carelessly handled by construction management, and frequently, by default, become a stepchild of the FCE. On reimbursable projects the owner's CEs frequently find that they must provide the impetus needed to guarantee the efficient handling of backcharges.

23

THE MONTHLY COST REPORT AND COST CONTROL MEETINGS

NEED FOR THE MONTHLY PROGRESS REPORT

Corporate management of many owner companies require monthly progress reports from managers of their large-capital projects, and the cost status, including forecasted total cost, is an important part of the progress report. Even when higher management does not require a monthly cost report, PMs need such reports for cost control purposes. The monthly cost forecast is the PM's tool for advising all concerned personnel of the cost impact of decisions and actions taken during the past month. Discussion of this forecast at monthly cost control meetings of the project management team is an effective way to focus attention on problem areas and to create a generally cost-conscious atmosphere.

The cost report, prepared by the CE, should include the following four major parts:

1. *Narrative section*: Highlights explaining any major changes to the cost forecast made during the past month. It also covers potential cost problems and corrective actions taken to minimize these problems. One should be able to read the narrative section and obtain a clear, concise picture of the current project cost status, including cost problem areas and action taken to minimize these problems.
2. *Project cost summary:* The major document in the cost report. It includes the following critical data for each cost code: budget, value of changes approved during the month, current dollar forecast, amount of budget over-/underruns, and value of commitments to date.
3. *Change order summary:* Includes a list of changes approved during the month, the amount of total changes approved to date, and a list of pending changes currently under consideration.

4.   *Charts and graphs:* Includes the critical productivity profiles, trend charts, and progress curves that are used by management for cost control of the project.

THE PROJECT COST SUMMARY

Although the narrative section summarizes the project cost status and keeps management up to date on the cost impact of current activities, the project cost summary (PCS) is the backbone of the cost report. This document gives the complete cost story in a format meaningful to PMs, engineers, and accountants.

A well-organized PCS will list the prime cost codes (cost elements) and provide the following data for each code:

Previous or last month's control budget
Change orders and budget shifts approved during the month*
Current control budget
Current cost forecast
Over- or underruns to the budget
Commitments and expenditures to date
Remarks

Every owner and contractor has their own format for presenting the cost summary and some of these omit important information. The common omissions are previous control budget and changes approved during the month. These omissions are tolerable providing the data are readily available elsewhere in the report. Occasionally one may find an old format still in use that includes only the original budget without changes, expenditures during the month, total expenditures to date, and commitments. Such a format is unsatisfactory for cost control and should not be tolerated on a reimbursable project.

Many contractors and owners present the monthly cost summary in a computerized form. This has many advantages in that a single data bank can be used by CEs who input the control budget and forecast and by accountants who input expenditures and commitments. The computer is an excellent tool for cost reporting. It can tell the CE how many manhours or dollars have been spent on each cost code and by whom. It can calculate productivities to the point of saying that to date foreman A poured concrete for 10 manhours per yd^3 and that foreman B averaged 13 manhours; however, it cannot tell why there was a difference or what next week's productivities are likely to be. For example, the computer cannot know, but the CE should know, that a crane made idle by

---

*Budget shift is the term used for shifting funds from one cost element to another. For example, a decision to subcontract an activity budgeted as direct labor would necessitate a budget shift.

late delivery of a heavy tower was used to help foreman A lift wet concrete into forms while foreman B had to use wheelbarrows and that the tower will arrive over the weekend and the crane will not be available to foreman A next week.

An owner's CE must be wary of a contractor's computerized report that includes a cost forecast. If the forecast has been worked up by hand and input separately it may be reliable, but if the forecast is calculated by a computer program it is suspect.

PROJECT COST SUMMARY: LEVELS OF DETAIL

Whether the PCS is produced by computer or manually, it normally consists of at least two and frequently three levels of detail. The most detailed level is broken down by detailed cost code; for example, painting might be divided into sand blast, apply prime coat, apply finish coat. This is the working-level document used by CEs and supervisors. The second-level summary report is broken down by prime code such as painting, insulation, and concrete. Many PMs include the summary-level cost report in their progress report; however, others prefer a management-level summary, which reports only basic cost elements, such as

Direct labor
Direct materials
Subcontracts
Engineering
FLOH
Other overheads

If a project is divided into two or more control areas, an added reporting complication is introduced. Complete detailed and summary-level reports are prepared for each area; however, all the areas are normally combined for the management-level report.

EXAMPLES OF COST SUMMARIES

Figure 1 is an example of the first page of a summary-level PCS used by a major engineering/construction contractor. This first page includes direct costs only: materials, labor, and subcontracts. The second page, not shown, includes the indirect costs: engineering, FLOH, ocean freight, etc. This type of summary is barely adequate for cost control. It breaks costs down by ten prime cost codes but it shows only four of the six desirable columns:

Current cost forecast (indicated total cost)
Current control budget (budget)
Commitments and expenditures
Remarks

Code	Cost element	Expended This month	Expended To date	Total commit	Cost forecast	Budget	+ Over – Under	Complete (%)	Remarks
	Direct labor								
0	Earthwork		497	497	497	450	+ 47	100	Over run in firewalls
1	Concrete	1	707	707	720	720		99	Fireproofing incomplete
2	Str. steel		188	188	188	195	– 7	100	
3	Buildings								
4	Equipment	80	342	342	350	327	+ 23	99	Compr. C-102 late deliv.
5	Piping	60	1422	1422	1530	1600	– 70	96	COs added $75 thousand
6	Electrical	26	491	491	560	570	– 10	90	
7	Instruments		180	180	310	297	+ 13	63	
8	Insul. and paint	51	283	283	505	495	+ 10	60	COs added $10 thousand
	Total	218	4110	4110	4660	4654	+ 6	90	4% gain this month
	Material								
0	Purch. earth		21	21	21	25	– 4		
1	Concrete	40	1160	1238	1254	1240	+ 14		
2	Str. steel		1062	1064	1064	995	+ 69		
3	Buildings								
4	Equipment	1252	14585	15043	15043	16881	– 1838		
5	Piping, valves	127	6567	6585	6590	6128	+ 462		COs added $322 thousand
6	Electrical	228	2934	2943	2950	3123	– 173		
7	Instruments	189	2812	2756	2860	2754	+ 106		
8	Insul. and paint	224	758	820	950	900	+ 50		COs added $30 thousand
	Total	2060	29899	30470	30732	32046	– 1314		
	Subcontracts								
3	Buildings	75	275	325	325	305	+ 20	90	
	Total	75	275	325	325	305	+ 20		
	Total direct	2353	34284	34905	35717	37005	– 1288		3.5% under budget

Figure 1 An example of the first page of a summary level PCS used by a major engineering/construction contractor.

Cost elements	Detailed control estimate	Previous control estimate	Budget shifts and changes	Current control estimate	Current outlook	Over + Under -	Commit to	Spent to
				\$ Thousand				
Materials	35,156	33,152	+ 124	33,276	30,869	-2,407	30,032	22,272
Labor	3,933	3,382	+ 58	3,440	5,983	+2,543	5,446	5,446
Subcont	4,476	10,434	-	10,434	9,325	-1,109	9,040	8,762
Total direct	43,565	46,968	+ 182	47,150	46,177	- 973	45,118	41,480
Indirects	30,432	32,404	+ 420	32,824	42,677	+9,653	35,866	35,882
Nonproject		13		13	205	+ 192	52	52
Undistrib CAs			+1,676	1,676	217	-1,459		
Target		79,385	+2,278	81,663	89,276	+7,613	81,036	77,214
Other costs	15,853	10,050	-2,179	7,871	8,591	+ 720	7,177	6,477
Credits	- 200	-1,117	- 911	-2,028	-1,867	+ 161	- 275	- 275
Total project	89,650	98,318	- 812	87,506	96,000	+8,494	87,938	83,416

Figure 2 An example of an overall management-level summary as used on a recent project.

It provides expenditures during the month and expenditures to date which are of interest to the accountant but not the CE, who generally works from the expenditure plus commitments column.

In actual practice Fig. 1 was run monthly on the computer with the indicated total cost and the over/under budget columns blank. The CEs then wrote in their forecast and sent the report back to the computer room where the entire report was rerun including all information as before plus the CEs' forecast in the indicated total cost column and the over/under budget column calculated. The result was a complete computerized report with the computer functioning as a combination typewriter and adding machine for forecast purposes.

Figure 2 is an example of an overall management-level summary as used on a recent project. This summary is too brief to be of any assistance in cost control, but it does provide corporate management with a quick synopsis of the project cost status.

MECHANICS OF REPORT PREPARATION

Information for the cost report originates in two places. The home engineering office provides cost data concerning engineering, bulk materials, equipment, inspection in vendors' shops, freight, insurance, and import duty if any. The construction office supplies costs for labor, FLOH, subcontracts, and field bulk material purchases. On small projects the engineering office is frequently in the same city as the construction site and one CE can serve both places. However, it is common for national engineering/construction contractors to perform the engineering and procurement for major projects in their home office, which may be hundreds or even thousands of miles from the construction site. As previously described, situations of this type require a team of CEs in the engineering office and another in the field, and preparation of a monthly project cost report requires close cooperation between the two groups. A few years ago the field construction report was either mailed or hand-carried to the home office for consolidation into the overall report. Now by the use of remote computer terminals, field data can be relayed to the home-office computer and the consolidated report can be printed out in both the home office and field.

THE FIELD COST REPORT

Regardless of how the information is relayed, the format in which it is tabulated, or where the consolidated report is prepared, the field report must contain the basic data shown in Fig. 3. This is a management summary-level report showing pertinent data for each of the basic cost elements controllable in the field. This report is, of course, backed up by a detailed-level report for each cost element. As we have seen, manhours is the focal point of field cost control, and for that

Cost code	Description	Costs						Manhours					
		Original budget	Approved changes	Current budget	Commit-ments	Estimated to complete	Forecast final	Original budget	Approved changes	Current budget	Actual man-hours	Estimated to complete	Forecast final	Progress percent
2000	Subcontracts (Material and Labor)													
3000	Subcontract labor													
4000	Direct hire labor													
5000	Temporary const: and consumables													
6000	Supervision													
6500	Tools													
7000	Payroll burden													
8000	Miscellaneous													
1000	Direct material purchased in field													
	Grand total													

Figure 3 A management summary-level report showing pertinent data for each of the basic cost elements controllable in the field.

273

reason the field report must show manhours as well as dollars for the labor-related cost elements including direct labor, temporary facilities, supervision, and miscellaneous indirect costs.

Figure 4 is an example of the field's summary-level manhours report as used by one contractor on a recent large project. In addition to the essential data previously mentioned, this example includes calculation of percent complete. It is a particularly good example of combining schedule and cost information on one page that can be used by all levels of construction management. It does omit dollar cost, which was provided on a separate page not shown here. This summary-level report was compiled by hand from computerized data except for the projection-at-completion (forecast) column. The projection was developed each month by the FCEs and reviewed by the craft superintendents before issuing the report.

CHANGE ORDER SUMMARY

The status of change orders is an important part of the PM's monthly progress report. The change order section usually consists of the following three parts:

1. A detailed list of changes approved during the month including the category, dollar cost, and schedule impact of each change.
2. A list of all pending changes including a magnitude estimate of cost and schedule impact for each change.
3. A graph showing the total value of approved and pending changes compared to contingency rundown.

CONTINGENCY MANAGEMENT

Funds to cover the cost of changes including estimate adjustments must come from one of three sources: project contingency, budget underruns, or additional appropriation. Almost all owners reserve additional appropriations for major scope changes only, and apparent underruns of some cost elements are usually required to offset overruns in others; therefore, it is customary to designate contingency or a specific portion of contingency to offset changes.

As changes are approved one can subtract the total value from contingency and show the remaining contingency in the forecast column of the PCS as the amount available to offset future changes. If this is done without control, one runs the risk of approving too many changes early in the project and leaving insufficient contingency to cover essential changes discovered later. To avoid this development, the CE must forecast how much contingency will be required to project completion and show that amount in the monthly cost report. This can be done by studying the rate of change order approval on past projects and developing a calibration curve exactly as done for the various field indirect

WORKHOURS AND SCHEDULE STATUS

Code	Description	Weight[a] factor	Physical % completion	Weight %	Workhours in 1000s				Schedule		
					Actual budget	QAB	Forecast	Total to date	1975	1976	1977
0	Earthwork	.08	97	7.76	505	505	477	450			
1	Concrete	.17	95	16.15	897	995	952	857			
2	Str. steel	.05	97	4.85	246	292	250	226			
3	Buildings	.00	97	0.00	64	64	68	65			
4	Equipment	.13	90	11.70	839	839	835	751			
5	Piping	.30	58	17.40	1773	1828	1900	955			
6	Electrical	.10	58	5.80	584	647	685	377			
7	Instruments	.07	45	3.15	445	445	514	222			
8	Insulation	.07	34	2.38	475	434	460	168			
9	Paint	.03	45	1.35	190	178	195	84			
9.1	Rework				79	82	90	63			
9.2	Startup assist.						69	10			
9.3	Back charges				69	69	70	52			
	Total	1.00		70.54	6166	6378	6565	4280			

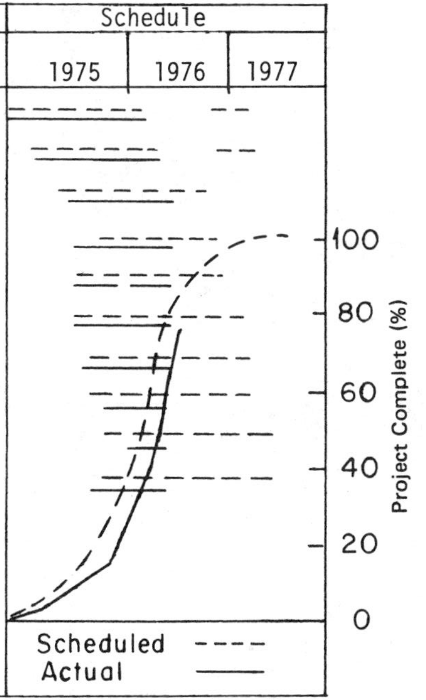

% Complete this month 70.54
% Complete last month 65.42
% Progress this month[b] 5.12
Productivity to date 1.05
Forecast productivity at completion 0.98

[a] Weight factor is based on QAB less rework and startup assistance.
[b] Productivity excluding backcharges and startup assistance

$$P = \frac{70.54}{(4280-10-52)/(6378-69)} = 105$$

Scheduled -----
Actual ———

Project Complete (%): 100, 80, 60, 40, 20, 0

Figure 4   An example of the field's summary-level manhours report as used by one contractor on a recent large project.

275

accounts. In the author's experience the rate of approving change orders (or the rate of contingency use) approximates a straight line from the beginning of detailed design through mechanical completion in the field. Figure 5 is an example of a contingency trend curve where actual change order approvals are plotted and compared to the theoretical or calibration curve.

Figure 5 illustrates a contingency management plan for a project with a total of $8.4 million of contingency. $1 million has been reserved for essential startup changes discovered after mechanical completion, and $1.3 million has been set aside to cover estimating errors and omissions. The remaining amount is available to offset project execution and design-development-type changes. The three divisions of contingency as well as the total amount of contingency are based on statistics from past project. A graph similar to Fig. 5 should always be included as part of the change order summary in the monthly cost report.

CHARTS AND GRAPHS

Specific graphs to be included in this fourth section of the cost report depends on the desires of the PM; however, most PMs are anxious to have as much data

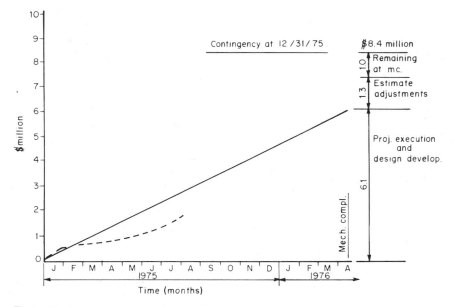

Figure 5   Contingency management.

as possible presented in an easily understood graphic fashion. As a minimum, consideration should be given to inclusion of the following:

1. Actual engineering percent complete compared to the scheduled percent complete: This curve has time as the abscissa and percent complete as the ordinate.
2. Procurement percent complete compared to the schedule: This curve is similar to the engineering curve with time as the abscissa and the percent of materials procured as the ordinate (percentage of materials calculated on a dollar basis).
3. Actual construction percent complete compared to the scheduled percent complete.
4. Labor productivity profiles at the prime code level for installation of bulk materials (i.e., concrete, structural steel, piping, etc.).
5. Tracking curves for the principal FLOH accounts including the following:
    a. Temporary facilities
    b. Construction equipment
    c. Supervision
    d. Material handling

BACKUP TABLES

Two important backup tables are essential to a complete report. The first of these is a list of all budget shifts (refer to Chap. 18) made during the month including a verbal description and the amount of money for each shift. This table can be retained by the FCE available for discussion with the PM. The second table is a detailed reconciliation of the current cost forecast with the preceding forecast. It lists all revisions made to the forecast during the month and can be included as part of the narrative section or annexed to the PCS.

MONTHLY COST MEETINGS

One of the most effective means of stimulating cost control is a monthly meeting. Ideally the meeting should be held after the monthly commited data are available and the CE has developed tentative cost forecasts. The meeting should be attended by all responsible project management personnel, and the purpose should be to review the cost impact of decisions and actions either taken during the month or contemplated in the near future. A prime concern of this meeting should be the consideration and evaluation of corrective action to minimize cost overruns. Meetings of this type will infuse a cost-conscious atmosphere into the project management team and provide the CE with information needed to finalize the forecasts and write the narrative section of the monthly report.

SUMMARY

In preparing the monthly cost report the CE must keep in mind the desires of the PM and the requirements of corporate management. The four sections described in this chapter are essential to give the complete cost status picture. The narrative section, probably the most important, is frequently the most slighted. Managers do not have the time to fumble through reams of paper covered by hundreds of closely printed numbers. They need a concise statement covering the important cost developments that occurred during the month and a clear picture of how these developments impact the outlook for total project costs. This information can be included in the narrative section and all else can be considered as backup data to be used by PMs and their staff in directly controlling the project.

24

THE COST OF COST CONTROL

INTRODUCTION

CEs demand that all aspects of an engineering project be justified. They ask, "How much does it cost and what is the payout?" It is only fair that they be judged by the same standards with which they judge. The cost of cost control is a part of the total project cost, and if that cost is greater than the benefits derived, then cost control should be eliminated. However, the facts are that cost control is a fast-growing field and payout has been proven on many projects. In this chapter we will examine a contractor's cost control organization, determine the cost of that organization, and compare that cost to the benefits received.

The cost of cost control can be divided into the following three distinct categories:

1. Cost of preparing a control estimate
2. Expenses of cost engineering during conceptual design
3. Expenses of the cost engineering group during detailed engineering and construction.

In the following paragraphs we will consider each of these categories individually.

COST OF A CONTROL ESTIMATE

Definitive estimates have been prepared ever since we have had construction contractors. In a capitalist society every owner, even the government, needs a cost estimate before initiating a construction project. Long before the term "cost control" was invented, detailed definitive estimates were required for the following reasons:

To establish a fixed price for lump sum bids
To determine economics and payout
To arrange for financing

Cost control required only that the estimate be prepared in a standard format corresponding to a code of accounts. This is normally done anyway because estimates are based on past cost data that have been collected by cost code; in other words, cost data are collected in the same way that money is spent. Detailed estimates are prepared in the same manner and an estimate so prepared is adequate for cost control. Whether we have a formal cost control program or not has little influence on the cost of preparing a detailed estimate. The estimate will be prepared for other reasons, and the incremental cost of using it as a cost control tool is negligible.

EXPENSES DURING CONCEPTUAL DESIGN

Estimating the cost of design alternatives has always been a feature of conceptual engineering. In the past, the designer frequently made these comparative estimates based on private "bottom-drawer" data collected over the years. This was expensive for two reasons: first, because highly skilled designers were spending their time estimating, and second, because the designer's incomplete data frequently gave the wrong result. It is normally economical to have design alternates estimated by a cost estimator. Since an estimator is required to estimate design alternates, this same estimator can act as the CE during conceptual engineering.

The formal reporting and trending described in Chap. 3 requires only an incremental amount of the estimator's time. In many cases it consists of logging the accepted design alternate if that alternate differs from the basis of the early estimate. For example, a project requiring 3 months of conceptual engineering might ordinarily require an estimator for two months. With a full cost-monitoring program the estimator would be required for all 3 months. The incremental difference would be only 1 month, say, $5,000 including salary, supervision, and overheads. Even on a small project, this would be considered next-to-negligible cost and the potential for cost reduction enormous.

EXPENSES DURING PROJECT EXECUTION

Cost control during project execution, the detailed-engineering and construction phases, requires sizeable expenditures. Figure 1 is an organization chart showing a contractor's cost control group used on a project requiring a total of 1 million engineering manhours. The home office or detailed-engineering group, including the principal CE, worked a total of 141 workmonths. D.J. Coster assisted in the estimation of change authorizations for approximately 6 months during the

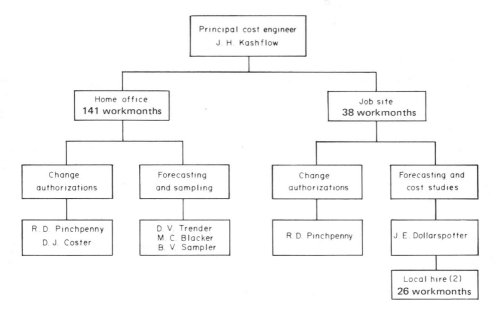

Figure 1   South Pole refinery organization chart, cost control group contract 0004.

period of model reviews, which resulted in a peak load of changes. Also, B.V. Sampler worked for approximately 6 months during the peak period of bulk design. The others were active throughout the duration of the project and R.D. Pinchpenny relocated to the field where she handled the estimation of field changes during the latter part of the project. She also assisted in forecasting direct labor. Dollarspotter accompanied the construction manager to the field and after mechanical completion he returned to the home office and assisted Trender in preparation of the final cost report. In addition to the contractor's cost group shown on the chart, the owner had one CE who was resident in the contractor's office during detailed engineering and who relocated to the field when engineering was about 90% complete.

Table 1 is a breakdown of the actual cost incurred by the contractor's cost group. Long before formal cost control systems were initiated a certain amount of estimating to determine economical alternatives and to support scheduling was done in both the home office and the field. Therefore, the total incurred cost, $362,000, has been reduced by $26,000, the estimated value of normal work without a cost-control program. The remainder, $336,000,

Table 1 Cost of Cost Engineering

Item	Work-months	Man-hours	$ Per manhour	Total $
Home office	141	24,400	10	244,000
Job site	38	6,600	13	86,000
Job site clerks	26	4,500	5	23,000
Expenses				9,000
Total		35,500		362,000
Less office estimate	10	1,700	10	−17,000
Less field estimate	4	700	13	− 9,000
Net cost engineering		33,100		336,000
% Of total engineering		3.8		3.9
% Of total project cost				0.4

represents the incremental cost of cost control and in this example amounted to 3.9% of total engineering cost or approximately 0.4% of the total project costs.

TANGIBLE BENEFITS OF COST CONTROL

In the authors' experience the tangible benefits of cost control (i.e., those hard-dollar savings picked up by the CE through carrying out normal duties) always exceed the cost of the CE group by a factor of 2 to 3. The intangible benefits of cost control are, therefore, obtained free of charge or at less than cost. Table 2 is a list of tangible cost savings directly attributable to the cost control group described in the preceding paragraphs.

OTHER EXAMPLES

Although the preceding example was taken from a project of approximately one million engineering manhours completed in 1970, it is still typical of current projects both large and small. A project with two and one half times the engineering manhours of the example and completed in 1982 spent approximately 0.5% of the total cost on cost control. On this large project, plagued with quantity, price, and manhour overruns, the contractor's cost engineering staff in the home office peaked at 22 persons and the field peak was 14 persons. These peaks were of short duration and occurred at times when the owner was deeply involved in reassessing the economics of the project.

From the preceding examples and others where the programs outlined in this text have been used, the authors conclude that good cost control requires an expenditure ranging from 0.35% to 0.5% of total project cost. When cost

Table 2　Examples of Saving Due to Routine Cost Control Activities.

Sampling of a group of computer-designed concrete footings showed that the volume of all small footings averaged between two and three times the budget estimate. Investigations revealed that all footings were set 4 ft. below grade regardless of location or loading. Correction of this resulted in the saving of a considerable amount of concrete, forming, and excavation.

A sampling of pipe supports pointed up about a dozen bents that were 22 ft. wide in a location where 12-ft. bents were sufficient. Correction of this saved several tons of steel and some concrete.

A sampling of piping located a high-side error in material takeoff amounting to nearly 2 miles of 14-in. pipe. If it had not been for sampling, this pipe would have been purchased and left as surplus at the end of the project.

A review of alloy piping bid tabs showed that, because of the quantities required, 12 chromium pipe cost more per ton than 304 stanless. As a result, 12 chromium pie was eliminated from the job for a savings of $18,000.

Comparison of the budget vs. bids for surface condensers revealed that the budget basis of epoxy-lined water boxes was lower cost than the bid basis of Cu/Ni lining. Alternate bids verified that appreciable savings could be realized by going epoxy. In this case the final decision favored Cu/Ni, but the budget comparison clearly revealed a reasonable lower first-cost route that might have been followed.

Comparison of expenditures for local field staff with the budget triggered an investigation that resulted in the reduction of 18 persons and a total savings of 157 workmonths equivalent to $95,000. A similar investigation into the methods of distributing ice water for drinking resulted in an estimated savings of $20,000.

Plotting of the cost trend curve for rented construction equipment highlighted excessive rental ($8,180 per month each) paid for two locally rented 70-ton cranes. After investigation it was discovered that these cranes could be released more than a month earlier than planned and the few remaining heavy lifts handled by moving an existing large crane from area to area as required.

A sharp drop in the monthly productivity curve for concrete labor instigated an investigation resulting in immediate reduction of 25% of the concrete labor force.

control exceeds this limit, it usually is the fault of the reporting system. Some companies have such complex reporting systems that the CEs spend full-time just trying to make the system work, and there is no time left for analysis. An example coming close to this is the performance measurement reporting that some U.S. government agencies impose on their contractors.

Table 3   Benefits of Cost Control

Savings effected by all persons on the project being cost conscious
Cost reductions made by focusing attention on trouble spots while there is still
     time to take corrective action
Deletion of changes that cannot be economically justified
Early notification of cost overruns that cannot be corrected (i.e., elimination
     of cost blackout periods)

The cost reporting system should be kept simple and accurate—the analysis should be thorough with reliance placed on charts and graphs. Reports to the project manager should be concise, focusing attention on the cost problems with underlying reasons explained along with recommendations for corrective action. One shrewd PM stated it all in a nutshell when he said: "I would like to see one-half the reports and twice the analysis."

INTANGIBLE BENEFITS OF COST CONTROL

The major benefits derived from cost control are not the tangible-type savings listed in Table 2 but the intangible savings as listed in Table 3. Referring to Table 3, no person familiar with the development of engineering/construction projects would seriously dispute the enormous potential for savings inherent in the first two points, and the CE through normal activities is instrumental in creating the cost-conscious atmosphere and focusing attention on trouble spots. The last point in Table 3 is probably the most important of all. The elimination of cost blackout periods gives management an added dimension of flexibility not available without cost control. This flexibility alone can justify the cost of cost control.

Personnel associated with cost control but not included in the preceding discussion includes timekeepers and those who measure progress. The data collected by these people are used by the CE, and frequently supervision of this data collection is part of the CE's duties. However, this information is required for scheduling and will be collected whether or not there is a formal cost control program; therefore, the cost of the data collection cannot be considered as an incremental expense attributable to cost control.

25

SELECTING A CONTRACTOR FOR
A REIMBURSABLE-TYPE CONTRACT

THE PROBLEM

In selecting a contractor for either a lump sum or reimbursable-type contract one must first ascertain the technical competence and general reliability of the various possible contractors. Determination of technical competence may involve a visit to the contractor's home office and interviews with key technical personnel. A prime consideration at this point is an evaluation of the contractor's existing workload and capacity to handle the projected additional work. Having ascertained competence, reliability, capacity, and willingness to bid, an owner is then faced with a choice based on economics. For well-defined projects where contractors are willing to bid lump sum, the choice is relatively simple, but for reimbursable projects the choice is frustratingly complicated. A contractor's typical proposal for reimbursable work will include the following:

A list of engineering positions and average salary for each position.
Actual names and short resumés of lead personnel who will be assigned to the project.
A list of construction staff positions and average salaries.
A list of construction equipment and current monthly rental rates.
Engineering overhead and payroll burden specified as dollars per manhour or as percentages of direct salaries.
A list of rates for miscellaneous expenses, such as dollars per minute for computer use and cents per copy for reproduction services.
Fees, preferably divided into engineering and construction portions. Engineering fee is frequently quoted as dollars per engineering manhour. Construction fee takes various forms, such as a percentage of payroll or dollars per supervisor manhour. However quoted, construction fees should always have a maximum limit.

285

A list of similar projects successfully completed by the contractor.
Most proposals will also include miscellaneous sketches, plot plans, schedules, and other technical data designed to impress the client.

Normally, reimbursable proposals do not include a cost estimate, but if they do, the estimate will be so studded with disclaimers that it cannot be relied upon. With only the meager cost data just described, an owner must select a contractor.

HOW TO ATTACK THE PROBLEM

By the use of estimating techniques we can evaluate the tangible portions of reimbursable bids. For example, estimate the engineering manhours and then apply the contractors' average salaries, overheads, burdens, and fees to these manhours. The result will be a comparatve engineering cost for each contractor. It is not necessary that the estimated base engineering manhours be accurate because we are not looking for absolute values but only for comparative numbers. The same technique can be used to compare the cost of construction staff, construction fee, and construction equipment. Following the preceding plan, we can make dollar comparisons of the tangible aspects of the various bids.

After evaluating the tangible aspects of each bid, we can turn our attention to the intangibles. For example, contractors who charge the highest rate per engineering manhour might employ more experienced engineers than competitors; therefore, they will use fewer manhours, prepare better designs, and overall, construct a lower cost project. To make comparisons of this nature we must have a subjective analysis of the technical qualifications of each contractor. We can then assign dollars to technical qualifications as follows: highly skilled designers can save 5% of the bulk material cost when compared to merely adequate designers; therefore, if one contractor is clearly more skilled than a competitor, we can credit him with a dollar savings of 1 to 5% (depending on the degree of superiority) of the estimated installed cost of bulk materials. If a contractor were incompetent in only one area, his bid might be debited with the estimated cost of an outside consultant or the cost of owner's technical personnel required to assist in the weak area. By applying the technique just described we can evaluate strengths and weaknesses of a contractor's engineering and construction organization on a dollar basis and apply this adjustment to our tangible evaluation of his commercial terms.

EVALUATION OF A CONTRACTOR'S COMPETENCE IN COST CONTROL

In reimbursable situations, a contractor's competence in cost control is important to the owner; therefore, it behooves an owner to develop a method of predicting the type of performance that he can expect from a particular contractor. A technique frequently used is to develop a list of pertinent factors and then rate the bidders on each factor using a scale of one to five. The contractor with the highest composite rating can then be judged as the most effective in cost control. Such a list of factors might include the following:

Status of the contractor's cost-control organization within overall organization. This can be determined by looking at the contractor's corporate organization chart, usually included as part of a reimbursable proposal. If cost control is not shown or if the head of cost control is shown reporting to the assistant accountant, one can be sure that this contractor's dedication to cost control is minimum. The manager of project control (cost scheduling, and estimating) should be on the same reporting level as the managers of engineering and procurement.

Evaluation of the contractor's written cost-control procedure. If a contractor is dedicated to the principles of cost control he will have a standard written procedure, and he will make it available to prospective clients. A good procedure goes beyond mere cost reporting and will emphasize analysis, corrective action, and follow-up. One should not be deceived by a procedure that starts out with a few pages of general "motherhood"—type statements followed up by a bulky, detailed description of the contractor's cost code.

Evaluation of the contractor's change authorization and cost-trending procedures. The contractor's written procedure should spell out a detailed and comprehensive cost-trending or monitoring program; also, it should cover estimating and handling of change orders.

Management philosophy and attitude toward cost control. If a contractor is commited to cost control, his prospective PM will emphasize the point during interviews with the client. He will be knowledgeable about control procedures and ready to answer questions about them.

Adequacy of the accounting and cost-reporting function. Methods of cost and progress reporting will be covered in the written cost-control procedures. Analysis and control are impossible without prompt, accurate reports; therefore, a prospective client must verify that the contractor's methods and reporting format are suitable.

Capability of the contractor to perform the required estimating functions. All successful contractors can prepare adequate detailed estimates although some may take longer and spend more money than others. However, contractors vary widely in their ability to make preliminary and semi-detailed estimates. If conceptual engineering is required of the contractor, estimating tools and experience are of prime importance.

Evaluation of the individual nominated to become the contractor's project CE. This individual should be interviewed and experience and qualifications determined. An unknown entity, without influence in the contractor's organization, should never be accepted for this position.

SUMMARY

The selection of an engineering/construction contractor for a reimbursable project is of considerable economic importance to an owner, and effort spent in selecting the most suitable contractor for a specific project will payout many times over. Proper selection involves the following steps:

1. Review the field of possible contractors and screen out those of lesser technical expertise and those who are already loaded to capacity.
2. Solicit proposals from the more likely and interested contractors.
3. Evaluate and compare the commercial terms of the bidders.
4. If the contractors are unknown, visit their offices and interview key personnel; then evaluate and compare the intangible aspects.
5. Interview and determine the capabilities and reputations of lead personnel proposed for your specific project.
6. Subjectively assign dollar values or weighting to the intangibles (steps 4 and 5) and add these to the tangibles, thus obtaining an overall rating for each contractor.
7. Award the project to the contractor with the best overall rating.

Part III

CASE STUDIES

CASE STUDIES

The following section is a collection of individual case studies which the authors have found beneficial in teaching the principles of cost engineering. The case studies are designed to expand the students' understanding of the fundamental principles discussed in the text. All of the studies depict real-life situations as they are encountered by practicing cost engineers, and like many real-life situations, there is not always a 100% right or wrong answer.

The first study can be worked after studying Chaps. 1-3. The next three studies require an understanding of the principles of estimating and tool making which are covered in the first ten chapters of the text. The remaining studies are illustrative of cost-control problems encountered in project management and can be worked progressively as the student studies Chaps. 11-25.

All of the studies can be worked individually by each student; however, the authors have found that some of the studies are best worked by teams of two to four persons, and these studies are so indicated. In real life most problems are solved or decisions are reached by review and discussion with other members of the project management team; therefore, the authors highly recommend this approach on the case studies so indicated.

The authors would like to thank Mr. Fumio Otsu of Fluor Engineers and Constructors for his contributions in preparation of the case studies.

Solutions to the case studies can be obtained from the publisher as a separate booklet.

CASE STUDY NO. 1: ORDER-OF-MAGNITUDE ESTIMATE FOR THE SECOND FILTER DRUM SUPPORT STRUCTURE

This case study should be solved after studying Chaps. 1-3.

Scenario

The date is December 1982. You are an estimator for the Ewing Oil Company in Dallas, Texas and have been requested to prepare an order-of-magnitude (Class D) estimate to assist the company in its evaluation of the economics of adding a second filter drum (FD-102) to one of its processes. Others have obtained a quotation for the second filter drum, and prepared estimates for all other associated items, except for the estimate of the cost of the structure to support the second filter. You are to estimate this last item and have been allocated only 60 min. to finish it.

Known Data

1. Design details for first filter structure (See Table 1).
2. Final quantities for first filter structure (See Table 2).
3. Final cost details for first filter (1977 costs) (See Table 2).
4. Following cost indexes (1977 to year end-1982):
    a. Structural steel material   =   1.27
    b. Steel fabrication   =   1.17
    c. Steel erection (S/C)   =   1.15
    d. Fireproofing material   =   1.20
    e. Fireproofing installation (S/C) =   1.10
    f. Overall factor   =   1.19
5. Escalation (steel and fireproofing)
    a. First year (1983) — M = 5%, Fab = 6%, L = 7%
6. Rough schedule —

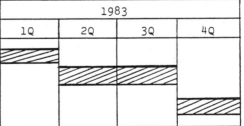

7. Design basis for second filter structure (see Table 1). Same as first filter structure except operations has reviewed the original design and has added another stairway on the west side of the platform. Also, recent changes in the safety standards require fireproofing of the top level (14 meter elevation).

Table 1 Design Details for First Filter Structure.

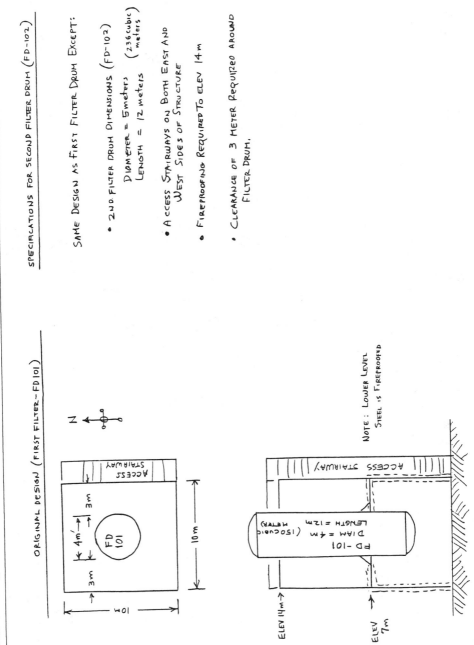

Table 2 Final Cost Tabulation for FD-101 (1977 Costs)

Item	Quantity	Unit Costs			Final Cost – 1977 (US$)		
		M	F	L	Material	Fabrication	Labor
Lower level of steel (to elev. 7 m)	12 tons	400	450	250	4,800	5,400	3,000
Second level of steel (to elev. 14 m)	8 tons	400	600	300	3,200	4,800	2,400
Grating support steel and handrails	4 tons	550	700	350	2,200	2,800	1,400
Grating (flooring)	6 tons (150 sq. m.)	1700	700	200	10,200	4,200	1,200
Stairway	3.5 tons	750	1200	225	2,600	4,200	800
Subtotal for Structural Steel					23,000	21,400	8,800
Fireproofing	20 sq.m.	200	–	900	4,000		18,000
Total Direct Costs (1977)					27,000	21,400	26,800

Problem

1. Prepare a Class D estimate of the material, fabrication and labor cost of the steel structure and fireproofing, within the time constraint mentioned earlier.
2. List and describe other approaches you could have used to prepare the estimate, other than the one you did use. Why didn't you use these?
3. Very briefly, describe how you could develop a simple estimating method from the data available for use in preparing future estimates for similar steel structures.

CASE STUDY NO. 2: SEMIDETAILED ESTIMATE FOR TIJUANA OIL CO. — NEW FEED PREHEAT TRAIN

This case study should be solved after studying Chaps. 1-10. The study covers direct costs only and comprises nine separate cost codes. The student may wish to partially solve the study by preparing estimates for only some of the cost codes. As a minimum the student should prepare estimates for codes 10, 50, and 80.

Scenario

You are the estimator in the cost engineering section of the engineering department of the Tijuana Oil Company. The process engineers have developed a plan for debottlenecking the existing kerosene hydrofiner. The main bottleneck is the feed preheat train which must be increased by 25% in capacity. To avoid a long plant shutdown, it has been decided to add a complete new feed preheat train. Other debottlenecking requirements along with tie-in of the new preheat train are minor and will be charged to expense and installed by refinery maintenance during a short plant shutdown scheduled for the fourth quarter of 1985.

The process engineers have been very late in developing this proposal; however, refinery management anticipates a rapid payout and wants to include it along with other budget proposals to be made to the board of directors next week. It is company policy that the board will only consider budget proposals based on *semidetailed estimates*. The head of the enigineering department has promised management that there will be no problem in developing the estimate since final costs of the exisiting preheat train (completed in 1976) are readily available and that cost engineers can develop a semidetailed estimate based on prorations within two days.

Problem

1. Your assignment—as the best estimator in the department—prepare the semidetailed estimate for all direct costs by cost code.

2.   List the principal areas of vulnerability (i.e., weakness) in your estimate.
3.   If more time were available, what would you do to improve the accuracy of your estimate?

Information Provided

Final cost report for the existing kero hydrofiner, feed preheat facilities.

Construction schedule for the existing kero hydrofiner.

Flow plans of the existing feed preheat train and of the new proposed train.

Partial materials list developed by process and mechanical engineering.

Plot plan of existing and new facilities developed by mechanical engineering.

Estimating guidelines extracted from Tijuana Oil's cost data book.

Proposed schedule for the new facilities.

Miscellaneous information obtained by questioning the process engineers and plant operators.

A copy of Tijuana's standard estimate summary sheet.

TIJUANA OIL COMPANY
FINAL COST REPORT - HERO HYDROFINER
FEED PREHEAT FACILITIES
Mechanical Completion   6/30/1976

| Code | Item | N⁰ | Unit | Work Hours | Final Cost in Dollars | | | |
					Labor	Mat'l	Sub.Cont	Total
10	**Equipment**							
	Feed Drum D-101	8	tn	100		9500		
	Pumps P-101A/B/C	60	hp	150		30600	(a)	
	Exchanger E-101	200	m²	80		10400		
	Sub Total			330	2480	50500		52980
20	**Structural Steel**							
	Columns/Supports	10	tn	100		8500		
	Grating Support & Handrails	4	tn	80		5000		
	Grating (flooring)	6	tn	90		10200		
	Stairway	2	tn	40		3900	(b)	
	Sub Total			310	2020	27600		29620
30	**Concrete**							
	Structure Footings	8	m³	160		760		
	Pump Footings	4.6	m³	90		440		
	Paving (15 cm)	110	m²	280		1570		
	Cable Trench	15	m³	80		450		
	Sub Total			610	3660	3220		6880
40	**Earthwork (c)**							
	For Footings	36	m³	140				
	Cable Trench	30	m³	120		150		
	Sub Total			260	1430	150		1580

(a) Includes motor drivers
(b) Includes fabrication cost
(c) Includes excavation & backfill

FINAL COST REPORT – HERO HYDROFINER
FEED PREHEAT   sheet 2 of 3

Code	Item	N°	Unit	Work Hours	Final Cost in Dollars			
					Labor	Mat'l	Sub. Cont	Total
50	Piping							
	Schedule 40							
	Line # Size Inches							
	1 8	35.2	m				2870	
	2A 8	10.0	m				820	
	2 10	18.3	m				2160	
	3A 6	10.0	m				620	
	3 8	18.3	m				1500	
	4 8	12.2	m				1000	
	5 6	27.4	m				1510	
	6 6	21.3	m				1320	
	Valves							
	6 inch gate	5	Nos				4600	
	8 " "	6	Nos				8400	
	6 " check	3	Nos				3000	
	Instrument Piping			100		850		
	Sub Total (a)			100	700	850	27800	29350
60	Electrical							
	Switchgear	3	pce	30		3700		
	Cable			110		1200		
	Conduit			100		400		
	Sub Total			240	1800	5300		7100
70	Instruments							
	Control Loops	2	Nos	100		10000		
	Misc Minor Items			50		750		
	Cont. Valves 6"	2	Nos	40		5000		
	Sub Total			190	1520	15750		17270

(a) Includes fittings & small valves , excludes control valves

Final COST REPORT - HERO HYDROFINER
FEED PRE HEAT   Sheet 3 of 3

Code	Item	Nº	Unit	Work Hours	Final Cost in Dollars			
					Labor	Mat'l	Sub.Cont.	Total
80	Insulation							
	Line # Size Inches							
	4 8	12.2	m				300	
	5 6	24.4	m				490	
	6 6	21.3	m				430	
	Exchanger E-101						800	
	Sub Total						2020	2020
90	Paint			160	960	500		1760
	TOTAL DIRECT COST			2200	14570	103870	29820	148260

Approved By: *D. Dollarspotter*
Comptroller

KERO HYDROFINER - ACTUAL SCHEDULE
Includes Feed Preheat Train

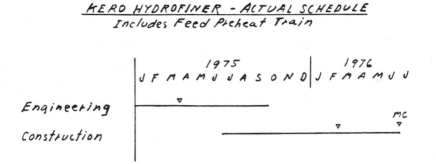

SCHEDULE FOR
PROPOSED FEED PREHEAT TRAIN

By: Tom Sharpshooter
Project Manager

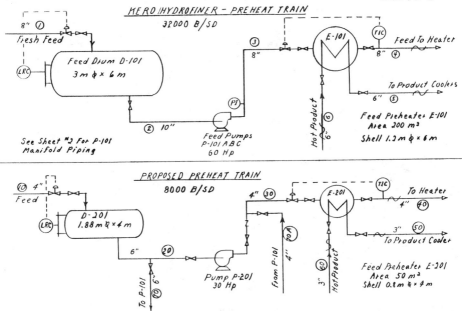

KERO/HYDROFINER – PREHEAT TRAIN
32000 B/SD

8" ①
Fresh Feed

LRC

Feed Drum D-101
3 m φ × 6 m

See Sheet No.2 For P-101
Manifold Piping

PI

② 10"

Feed Pumps
P-101 A B C
60 Hp

③
8"

E-101

TIC

Feed To Heater
8" ④

To Product Coolers
6" ⑤

Hot Product
6" ⑥

Feed Preheater E-101
Area 200 m²
Shell 1.2 m φ × 6 m

PROPOSED PREHEAT TRAIN
8000 B/SD

⑩ 4"
Feed

LRC

D-201
1.88 m φ × 4 m

6" ②⓪

To P-101 ⑪ 6"

Pump P-201
30 Hp

From P-101

4" ⑦⓪

4" ③⓪

E-201

TIC

To Heater
4" ④⓪

To Product Cooler
3" ⑤⓪

3" ⑥⓪
Hot Product

Feed Preheater E-201
Area 50 m²
Shell 0.8 m φ × 4 m

By A. Einstein   11/8/83
Chief Process Engr.

FEED PUMPS P-101 A B C
MANIFOLD PIPING

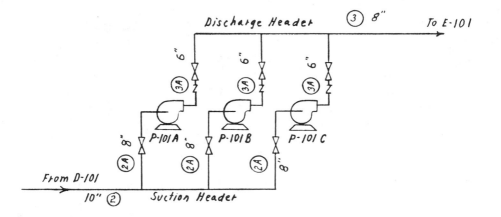

Discharge Header   ③ 8"   To E-101

6"   6"   6"

③A   ③A   ③A

P-101 A   P-101 B   P-101 C

8"   8"   8"

②A   ②A   ②A

From D-101

10" ②   Suction Header

<u>NEW FEED PREHEAT TRAIN FOR KERO HYDROFINER</u>

<u>PARTIAL MATERIALS LIST</u>

<u>Code 10, Equipment</u>

- o Drum D-201, 1.88 m dia by 4 m long, approx. weight 3 tons.
- o Pump P-201, centrifugal pump with motor driver, approx. 30 bhp.
- o Exchanger E-201, same materials as E-101, 50m^2 of heat ex-
     change area. Estimated shell size = 0.8 m dia x 4 m long.

<u>Code 50, Piping</u>

Line N$\underline{^o}$	Size Inches	Estimated Length in m	2" Insul- ation	Nos. of Valves Gate	Check
10	4	35.0	No	1	
20	6	9.0	No	1	
30	4	9.0	No	2	1
40	4	12.2	Yes	1	
50	3	24.4	Yes	1	
60	3	21.3	Yes	1	
70	6	5.0	No	1	
70A	4	5.0	No	2	1

<u>Misc. Other</u>

Other material quantities are minor and can be developed by
cost engineering.

By: R. Goldberg
Chief Mechanical Engineer

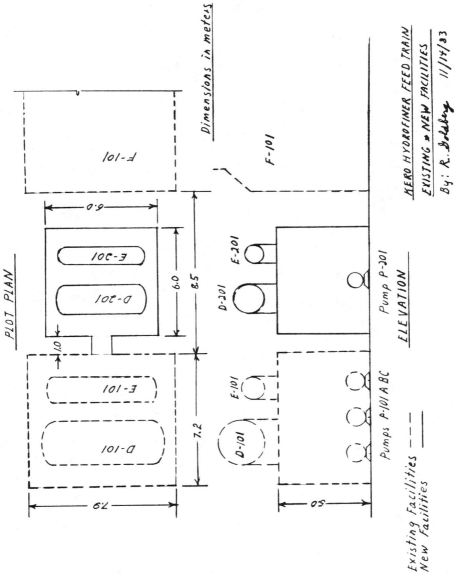

PLOT PLAN

Dimensions in meters

AERO HYDROFINER FEED TRAIN

EXISTING = NEW FACILITIES

By: R. Goldberg   11/14/83

ELEVATION

Pump P-301

Pumps P-101 ABC

Existing Facilities — —
New Facilities ———

ESTIMATING GUIDELINES EXTRACTED FROM TIJUANA'S COST DATA BOOK

Code 10 Equipment

Recommended slopes (exponents) for prorating equipment costs

Equipment	Parameter	Material exponent	Labor exponent
Towers & drums	weight	0.9	0.7
Heat exchangers	exchange area	0.8	0.7
Pumps & drivers	horse power	0.6	0.4
Motor switchgear	horse power	0.3	0.0
Turbine generator	MW	0.6	0.3

Escalation Index (at year end)

Year	Refinery material average	Labor
1973	136	150
1974	181	166
1975	189	178
1976	192	194
1977	199	210
1978	218	229
1979	239	247
1980	263	273
1981	302	300
1982	338	321
1983 (3Q)	355	331

Forecast Escalation Percent

Year	Refinery material average	Labor
1984	5%	4%
1985	6%	6%

Revamp Factors

Revamp class	Bulk Matls[a]	Direct labor	Field labor overheads	Engineering
A	5	10	15	25
B	8	16	25	40
C	10	20	35	50

[a] Excludes major equipment in Code 10.

Description of Classes

A: Plant addition constructed at end or side of exisiting plant. Ready access from three sides. Little or no conflict with existing piping, structures, or wiring. All work can be done independent of plant shutdown.

B: Plant addition constructed in reasonably open spaces between existing equipment. Access from two sides. Some conflict with existing piping, structures, or wiring. Majority of work can be done independent of plant shutdown.

C: New equipment scattered among existing equipment. No clear access. Considerable conflict with existing piping, structures, and wiring. Very little work can be done independent of plant shutdown.

Average Wages of Construction Workers — 3rd Quarter 1983

Cost code	Average wage dollars per hour[a]
10	13.73
20	11.84
30	10.90
40	10.16
50	12.80
60	13.58
70	14.55
80	10.50
90	10.78

[a] Feedback from a project currently under construction in refinery.

INFORMATION OBTAINED BY QUESTIONING
PROCESS ENGINEERS AND PLANT OPERATORS

1. The existing feed line to the fired heater (line #4) passes through the construction area. The plant will not shutdown to relocate this line — new construction must build around it.

2. New line #40 will tie into existing line #4 at the fired heater. This and all other tie-ins will be made by refinery maintenance forces during plant shutdown and charged to expense.

3. Operating temperatures and pressures in the new train will be the same as in the existing plant. New piping will be of the same materials and schedules as existing piping.

4. New instrumentation will duplicate the existing. There is space for the new instruments on the existing panel board in the control house.

5. There is capacity and space available in the existing electrical substation, but new low voltage switchgear will be required for pump P-201. Spare electrical conduit is available from existing junction boxes to both the substation and the control house. Existing junction boxes are about 4 m from new pump P-201.

6. Pump P-201 will not be spared but will be tied to the common spare P-101C via a jumpover, lines 70 and 70A. Since material is available, refinery maintenance forces will install line 70A as part of tie-in work.

7. A new access stairway is not required. There will be a short walkway between the new and existing platforms.

8. New insulation will be 2-in.-thick—the same as in the existing plant.

9. Dimensions of existing drum D-101 are 3—m-diam × 6-m-long. Dimensions of existing exchanger E-102 are 1.2m-diam × 6-m-long.

10. The original kero hydrofiner and feed train were erected by Lumor under a fully reimbursable fixed fee contract. Lumor is currently doing other work within the refinery and is familiar with existing equipment, safety requirements, and hot work permit procedures. Tom Sharpshooter has been approinted project manager for the new preheat train. Tom plans to engage Lumor to do both engineering and construction under a reimbursable contract. His reasons are as follows: First, to meet the scheduled shutdown, engineering must start immediately and there is no time to get competitive bids. Second, the majority of construction work must be done while the plant is operating with site access from two sides only, and Lumor is equiped to do this better than an outsider. Third, Tom feels that difficulties of working in an existing plant combined with lack of competitiveness will make a reimbursable contract less costly and easier to administer than lump sum.

CASE STUDY NO. 3: TOOL MAKING: NEW INVESTMENT CURVE FOR FEED PREHEAT FACILITIES

Case study number 2 provides background and should be solved, or at least reviewed, before attempting this study.

Scenario

You are a member of the cost engineering section in the engineering department of the Tijuana Oil Company. The estimate for the new feed preheat train described in case study number 2 has now been completed. The cost engineering section head, Pedro Ojo-Deaguila, has been advised that several new feed preheat trains are under study as part of an overall energy conservation drive, and he has requested you to prepare an investment curve to be included in Tijuana's cost data book.

Pedro formerly worked in Tijuana Oil's second refinery at Punto Aceite and he recalls that they constructed a 70,000 barrels per stream day (B/SD) pipestill in 1978 with feed preheat facilities identical to those described in case study number 2. Pedro has sent a telex to Punto Aceite requesting final cost data for these facilities.

Problem

Your problem is to prepare an investment curve for feed preheat facilities using information available from case study number 2, data obtained from Punto Aceite, and the following additional information.

Pedro suggests that the curve should be plotted on log-log paper using capacity in barrels per stream day (B/SD) as the cost sensitive parameter. A sheet of log-log paper is included for this purpose.

Additional Information

1. The total direct cost estimate for the proposed 8000 B/SD feed preheat train of case study number 2 is as follows:

	Estimated cost	Included escalation	Included revamp	Factors
Direct labor	$ 13960	10%	16%	
Material	93040	6%	8%[a]	
Subcontract	27060	8%	12%	
Total direct cost	$134060			

[a]On all materials except code 10 estimated at $29080.

2. The location of the facilities described in case study number 2 is used as standard location in Tijuana's cost data book. Base year for investment curves in the cost data book is 1980 with an average material index of 263 and a labor index of 273.

3. It is Tijuana's standard practice to plot investment curves reflecting total direct costs for on-site facilities. Average percentage splits for labor, material, and subcontracts are usually given as footnotes to the curves. It is understood that these splits are to be used only as rough guides in preparing early estimates. Off-sites and indirect costs are estimated separately and are not reflected in the investment curves. All of the curves reflect "grass roots" conditions.

4. A telex from Punto Aceite includes the following information on the feed preheat train completed in 1978:

    a. The facilities were constructed as part of a new "grass roots" atmosphere pipestill completed in December 1978. The centroid of construction manpower occurred about March 1978 and the centroid of materials procurement was approximately December 1976. Piping and insulation were subcontract and all other labor was direct-hire. The entire project was engineered and executed by Cieloalto Engineers and Constructors Ltd. under a fully reimbursable contract.

    b. Final direct costs were as follows:

Direct labor	$ 26300
Material	188500
Subcontract	52400
Total	$267200

5. Cost indexes from Tijuana's cost data book as given in problem number 3 are applicable.

6. For a number of years Punto Aceite labor productivity has been about 10% less than at Tijuana and piping erection subcontracts average 5% higher at Punto Aceite. The productivity difference has been attributed to more aggressive union activity at Punto Aceite. Wage rates and material costs are approximately the same at both locations.

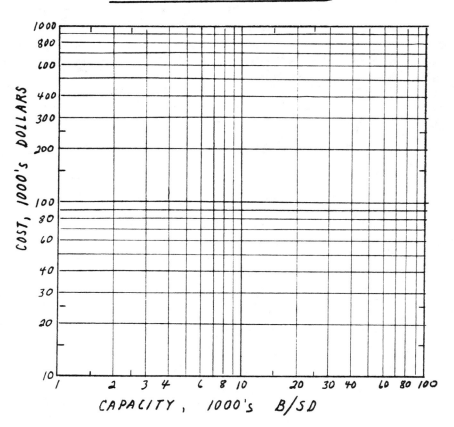

FEED PREHEAT FACILITIES

CASE STUDY NO. 4: CONCEPTUAL ENGINEERING: EVALUATION OF AN EARLY CONCEPTUAL ESTIMATE

Sooner or later every cost engineer gets involved in assessing the work of others. Corporate managements make economic decisions based on cost estimates, and when these estimates are prepared by third parties, they turn to their CEs to provide technical evaluation. This puts a heavy responsibility on the CEs because the decision to commit or not to commit large sums of money frequently hinges on the CE's assessment.

This cost study is designed to provide experience in a real-life situation where the CE is required to exercise judgement, based on estimating expertise, in making recommendations which will influence the future course and profitability of the company. It is suggested that after studying Chaps. 1-10 and completing the previous case studies, students be divided into groups of two to four persons to work this case study.

Scenario

You are an experienced estimator working for Ewing Oil Company which now desires to expand into North Sea offshore oil and gas exploration and production. Ewing currently has an opportunity to become a partner along with several other major oil companies in the development of a multibillion dollar offshore project. One of the other partners, Essell, is the operator for the project and has recently submitted to your management a "Declaration of Commerciality" (DOC) for the project. Your management has, in the past, expressed some misgivings about the project because of what they consider to be an abnormal number of uncertainties associated with the project, even for an offshore venture. Consequently, they have placed high priority and considerable emphasis on their request for an in-house assessment of the conclusions reached in the DOC (i.e., is the project really technically feasible and commercially viable, as indicated by the operator in the DOC?). A team or task force has been organized to provide the answer to management and you are a member of the team, with the prime responsibility of reviewing and commenting on the reliability of the capital investments used in determing the economics. Your company still has the option of withdrawing from the partnership if they deem the project to be unfeasible or uneconomical.

The Problem

Your problem, to be worked as a team, is to do the following:

1. Analyze the operator's cost estimate, estimate basis and approach and prepare a list of the questions you will want to ask him to allow you to complete your analysis.

2. From your preliminary analysis, what are the cost areas that have the highest degree of potential variability? Explain why.
3. Where do you feel the estimate is weak and why?
4. What, if anything, can be done to improve the estimate?
5. If you had been the operator's estimator, what would you have laid down as the minimum criteria for establishing a good basis for preparing the estimate?
6. What, in your opinion, is the accuracy of the estimate?

Information Provided

Summary description of the project
Estimate summary
Estimate basis summary
Risk contingency analyses

Summary Description of the Project

1. Major Facilities
   a. Offshore platform, with combined drilling, production, and living quarters.
   b. A network of six subsea installations, four two well and two single satellites.
   c. Interconnecting flow lines between the subsea installations and the "mother" platform.
   d. A 350 KM 20 in. diam. pipeline to shore.
   e. A major onshore terminal (by others).
2. Design Conditions
   a. Water depth of 350 m, twice the depth of any platform built to date.
   b. Poor sea bed soil conditions (soft clay).
   c. Unusually strong undersea currents.
   d. Topside facilities consist of 450,000 BPD of oil processing (no gas), the largest such installation to date.
3. Execution conditions
   a. Relatively tight 7-year overall schedule.
   b. Operator plans to use a considerable number of his resources and resources of one of the major partners to manage the job, along with contractor personnel, as required, to form an integrated management team. The major partner as well as several other oil companies have recently announced other development plans for roughly the same time frame.
   c. It is known from previous projects that there will be considerable emphasis placed by the government on the use of local materials and local resources.

d. Although the subsea installations are not a "first," they are the first to be built in this water depth, and under these sea conditions. The schedule includes some time for a comprehensive pilot and development program.

4. Estimate summary (all investments shown in local currency)

		M
a.	Structure (platform)	1,500
b.	Topside facilities	1,000
c.	Subsea installations	750
d.	Flow lines	500
e.	Pipeline	850
	Total project investment =	4,600

5. Estimate basis

a. All costs include escalation in accordance with the project schedule.

b. Structure costs represents the GBS* (local construction) but a comparable steel structure was estimated to be a standoff with, however, a six month schedule advantage. Both costs are based on quotations from the respective contractors.

c. Topside facilities are conventional, although large. Costs are based on a proration of the previous largest installation.

d. Subsea installations are based on the only previous installation, which only recently went into commercial operation.

e. Pipeline is conventional and costs are based on the operator's cost data bank.

f. Although not stated, it is assumed that all required contingencies and allowances are included in the individual numbers.

g. The operator did describe in the DOC a risk contingency analysis performed (using the Monte Carlo approach) which resulted in a 30% difference between the "most likely" number and the "high" number. The investment used in the DOC is the "high" numbers which the operator is presenting to your management as an investment which has a 90% probability of not being exceeded.

h. The "risk contingency" is described as covering the following:

1. Design changes within the scope of the work, limited in total to ±10 of the total investment (i.e., no major changes are covered).

2. "Normal" variations in the execution basis.

3. "Normal" estimating errors, omissions, and price deviations.

4. Abnormal weather, productivity, currency fluctuations, but is not meant to cover any "disaster" situations.

6. Cost comparision tables of such major items as the unit cost of concrete for the GBS, the unit cost per pound for fabrication and installation of the topside facilities, have been included as backup to show that the unit prices used are in line with previous projects.

*Gravity Base Structure, i.e., concrete.

CASE STUDY NO. 5: CONTROL DURING BASIC DESIGN: BID DOCUMENTS FOR BASIC DESIGN ENGINEERING

This case study is designed to test and expand the student's understanding of the principles discussed in Chap. 13 and to further acquaint him/her with real-life situations that confront practicing cost engineers. It is suggested that students be divided into groups of two to four to work this case study.

Scenario

You are now on loan from Ewing Oil and filling the position of senior cost control engineer for Essell, the same operator as in case study No. 4, and the field has now been declared commercial. Essell is preparing to go out for bids from contractors for a basic design engineering contract. You are a member of the bid team responsible for the contracting effort and for making the final recommendation to management. Your specific responsibility is to outline the minimum control requirements in the invitation-to-bid documents and to review the contractor's response. Most of the remainder of the bid review team consists of design and engineering personnel. In general, your management and your colleagues in the bid review team are inexperienced in project control and cost control in particular (they never took the opportunity to read *Applied Cost Engineering*!). You are aware from past experience that the bidding contractors (all local except Lumor) are weak in the controls area, but their international partners (they are bidding as joint-ventures) do have the capabilities and resources.

The Problem

1.   Prepare a detailed outline or listing of the minimum project control requirements which need to be included in the invitation-to-bid documents. In particular, emphasize the cost-control requirements and assume a colleague is working up the schedule, bulk material control, and sub-contract administration for you.

2.   Specifically, outline the approach you would expect the contractor to take, which procedures he should have and use, the project control organization, and reporting requirements.

3.   During the basic design phase, how critical is the contractor's cost control capability? How would you analyze his potential performance in this area prior to the award of the basic design engineering contract? What would you do with this analysis?

4.   What steps would you take and what arguments would you use to convince the other members of the bid review team and your management that all the information you are requesting from the bidding contractors is essential?

CASE STUDY NO. 6: ENGINEERING PROGRESS MEASUREMENT: EXPANSION OF MONACO-ENCON FACILITIES BY LUMOR

This case study should be worked after studying Chap. 15. It is designed to test the student's understanding of physical progress measurement during detailed engineering. It is a real-life situation that will eventually confront nearly all CEs engaged in project management.

Scenario

The Monaco-Encon company has engaged Lumor, a large international contractor, to engineer, procure materials, and construct a major "grassroots" addition to the existing oil refining facilities in Monaco. You are Monaco-Encon's cost engineer assigned to the project management team resident in Lumor's office.

In Lumor's latest progress report, detailed engineering was reported as 68% complete as of November 30, 198X. This was composed of a reported 67% contribution from the design and drafting area and 72% from the engineering specialist area.

Lumor's analysis of physical progress of engineering specialist work appears satisfactory and the reported 72% complete is judged as an accurate assessment. However, their system for measuring progress in design and drafting has been completely unacceptable and your repeated efforts to get Lumor to implement a more reliable system for assessing progress have been unsuccessful. As a result, you have decided to make an independent analysis of the status of design and drafting.

For each engineering discipline, Lumor's computerized progress report includes the numbers of drawings, control estimate manhours, expended manhours and estimated manhours to complete the work. A second report prepared by hand provides drawing status by the following four categories:

1.   Layout and general arrangement complete
2.   Construction drawing complete
3.   Construction drawing checked
4.   Corrections made and drawing issued for construction

Problem

Your assignment is to assess Lumor's reported status of overall detailed engineering. The following completion percentages have just been agreed as reasonable by both your project manager and Lumor's project manager and should be used to determine the status of the construction drawings.

	Percent Complete
Layout and general arrangement complete	30
Construction drawing complete	70
Construction drawing checked	90
Check corrections made, drawing issued for construction	100

Information Provided

Pertinent data from Lumor's computerized progress report and from the drawing status report have been consolidated on the attached worksheet.

Monaco Encon Project
Design & Drafting Status
Calculation Sheet

Date: December 1, 198X

Functional area	Description	Quantity	Manhours			Drawing status				Physical	Progress	Status
			Control estimate	Expended	To complete	Layed out	Drawn	Checked	Issued for construction	Percent complete	Weighting	D/D Completed
	Data from Lumor's progress report and drawing status reports											
P & ID'S:	Piping and instrument diagrams	15	4,500	4,870	460	15	15	13	11			
Civil:	Plot plans	3	1,200	1,470		3	3	3	3			
	Foundations	18	1,750	1,890		18	18	18	18			
	Structural	52	4,225	4,150	1,175	44	41	32	28			
	Structures (steel)	2	300	310	130	2	2	2				
	Miscellaneous (paving, grading, etc.)	4	500	380	320	3	3	2	1			
Piping:	A/G piping plan	17	6,800	6,180	1,240	17	15	13	12			
	U/G piping plan	6	600	580	160	6	5	5	4			
	Isometrics	1,050	21,000	9,420	11,580	615	430	240	110			
	Standards	18	1,800	1,510	690	17	15	10	8			
	Miscellaneous	20	2,000	860	1,140	8	6	5	3			
Pressure vessels and tankage:	Towers	10	1,500	1,340	710	8	6	6	5			
	Drums	7	460	390	170	6	6	5	3			
	Tanks	1	100	140		1	1	1	1			
	Drier	1	100	130		1	1	1	1			
Electrical:	Area classification drawings	1	240	270	60	1	1	1				
	One line diagrams	1	300	200	140	1	1					
	Cable schedules	9	450	110	340	4	2					
	U/G cable layouts	11	1,100	960	540	9	7	4	4			
	A/G cable layouts	4	400	90	310	1	1	1				
	Lighting layouts	4	400		400							
	Grounding plans	4	400		400							
Instrumentation:	Index & location plans	4	600	140	460	2	1					
	A/G inst. wiring & tubing racks	10	1,000	100	900	4	1					
	Instrument hook-ups	40	400	80	320	13	8					
	Panel board layouts	4	1,200	360	840	2	1					
	Control house layouts	3	900	440	460	2	1					
	Total	1,319	54,225	36,370	22,945	802	590	358	203			

Note: D/D = 74.2% of detailed engineering
E.S. = 25.8% of detailed engineering

CASE STUDY NO. 7: SAMPLING OF BULK MATERIAL DESIGN QUANTITIES: EXPANSION OF MONACO–ENCON FACILITIES BY LUMOR

This case study should be undertaken after studying Chap. 17. It is a real-life situation included here to improve the student's ability to analyze and make recommendations based on the results obtained from bulk material sampling. It is suggested that students be divided into groups of two to four to work this case study.

Scenario

Refering to cost study No. 6 you are still acting as Monaco-Encon's cost engineer resident in Lumor's engineering office. The status of the project is P&ID's* are essentially complete; piping routings and layouts are well underway; and general arrangements and isometrics are about 10% complete overall. The total number of isometrics for the projects is still estimated to be about 1,000 with sufficient drawings released (5%) or drawn (15%) to allow representative sampling. We have been unsuccessful in getting Lumor to get the sampling started so we have taken on the job to do a preliminary sampling. One typical isometric and two general arrangement drawings are attached for this purpose as well as the results of another 17 lines which we have sampled.

Problem

Your assignment, using the attached excerpts from the control estimate, the attached drawings, and results of other lines sampled, is to complete the following:

1. Appraise the contractor's piping design performance on lines 6"-0H02 and 10"-0H01 using the attached sketches SK-1, SK-2, and SK-4. The following abbreviations are used on the sketches: BOP = bottom of pipe, EL = elevation above grade, FF = flange face.
2. The results of the samples are summarized on the attached sample results worksheet. Complete this worksheet indicating projected over/underrun in quantities.
3. List the possible reasons for the over or underrun.
4. List possible corrective actions for the base case, as described and for the situation if piping design was 65% complete instead of 10%.
5. If corrective action is not taken or could not be taken, indicate on Table 1 which of the execution areas will be affected and provide an estimate of the impact on cost and schedule for each of these areas (i.e., none, slight, significant, or major).
6. If the results of the sampling analysis indicated an overrun in quantities, explain how best to approach and convince the contractor that he has a

*Piping & Instrument Drawings.

serious problem. (This assumes the contractor has not done the sampling or has not carried out a thorough analysis of the results.)

7. Should the contractor delay any extensive investigation and analysis of the quantity deviation until after receipt of material unit price bids and make one combined analysis based on the total impact of quantities and unit prices? Give reasons to support your answer.

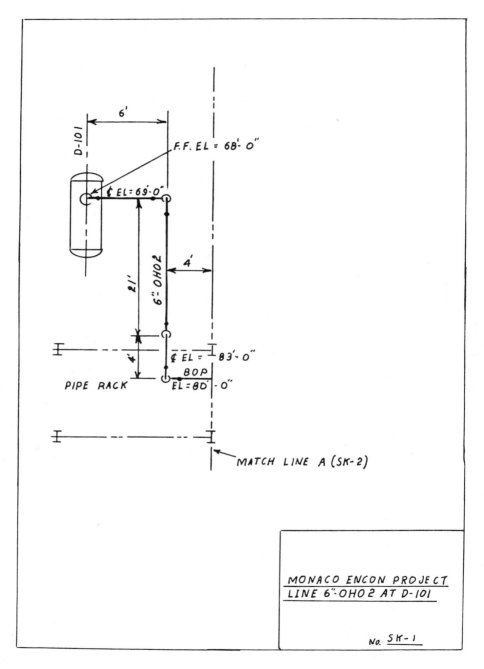

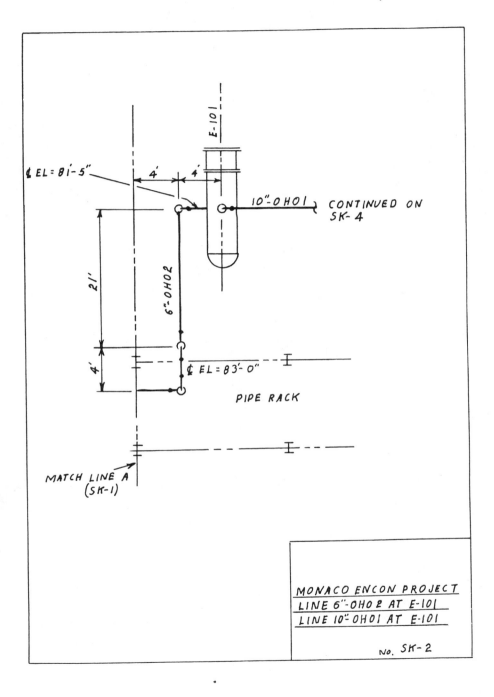

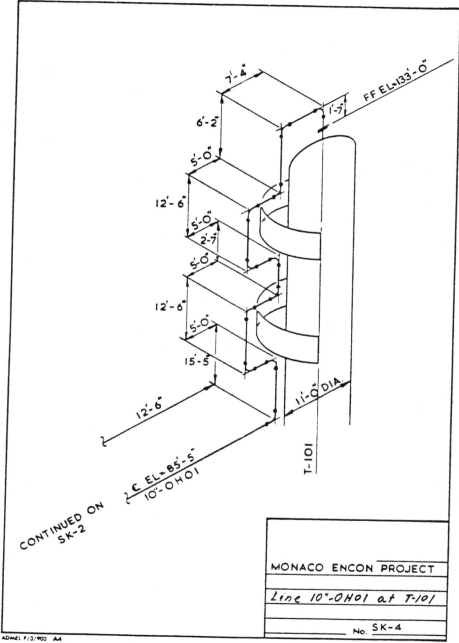

7'-4"

6'-2"

FF EL=133'-0"

1'-7"

5'-0"

12'-6"

5'-0"

2'-7"

5'-0"

12'-6"

5'-0"

15'-5"

11'-0" DIA.

12'-6"

T-101

₵ EL=85'-5"
10"-OH01

CONTINUED ON
SK-2

MONACO ENCON PROJECT

Line 10"-OH01 at T-101

No. SK-4

ADMEL P/3/903 A4

MONACO ENCON PROJECT COST/SCHEDULE SAMPLE RESULTS

Line No.	Feet of Pipe		Fittings		Flanges		Valves	
	Actual	Budget	Actual	Budget	Actual	Budget	Actual	Budget
2	430	365	18	14	0	0	0	0
3	495	348	10	10	0	0	0	0
4	258	220	12	12	0	0	0	0
6	1535	1345	48	46	1	1	2	2
7	1557	1254	50	42	1	1	2	2
8	183	160	10	4	0	0	0	0
61	855	595	32	32	0	0	0	0
63	422	375	13	13	0	0	0	0
64	422	375	13	13	0	0	0	0
80	159	140	10	4	0	0	0	0
89	56	40	8	8	4	4	1	1
91	56	40	8	8	4	4	1	1
93	56	40	8	8	4	4	1	1

MONACO-ENCON PROJECT COST/SCHEDULE SAMPLE RESULTS (cont.)

Line No.	Feet of Pipe		Fittings		Flanges		Valves	
	Actual	Budget	Actual	Budget	Actual	Budget	Actual	Budget
101	108	76	12	10	8	7	4	4
103	26	23	10	4	2	2	1	1
106	128	115	14	8	2	2	0	0
108	12	10	14	4	1	1	0	0
OH-01*		62		4		2		0
OH-02*		30		2		2		0
Totals		5613		246		30		12
Δ(Actual - Budget)								
Percent overrun (+) or underrun (−) (Δ/Budget) × 100%								
Total quantities for project (From Budget) (tons)	117.0		10.1					
Projected quantities required (tons)								

*Actual quantities for Lines 6 - OH-01 and 10 - OH-02 to be developed from sketches SK-1 and SK-2.

Table 1   Monaco-Encon Project Cost/Schedule Execution Area Effects

	Impacts			
Execution area	None	Slight	Significant	Major
Detailed engineering				
Procurement (purchasing)				
Expediting				
Inspection				
Material costs:				
Piping				
Pipe supports				
Concrete (foundations)				
Scaffolding				
Structural steel				
Instruments				
Electrical				
Insulation				
Paint				
Steam tracing				
Others				
Labor costs for:				
Piping				
Pipe supports				
Concrete (foundations)				
Scaffolding				
Structural steel				
Instruments				
Electrical				
Insulation				
Paint				
Steam tracing				
Others				
Field labor overhead costs:				
Supervision				
Tools				
Temporary construction & consumables				
Payroll burden				
Other				
Fees				
Others				

CASE STUDY NO. 8: CHANGE ORDER CONTROL: UNUSUALLY HIGH NUMBER OF REQUESTS FOR CHANGES

This case study should be solved after reading Chap. 18. It depicts a problem that is frequently faced by project managers and cost engineers—both owners and contractors. It is suggested that students be divided into groups of two to four persons to work this case study.

Scenario

You are a contractor's cost engineer on a large project for Monaco-Encon, Inc. involving major modifications to an existing plant. Engineering and procurement is well underway (65% complete) and constuction is about 15% complete. One of the more significant problems which has developed during engineering and appears to also be developing in the field is an unusually high number of requests for changes, coming from all sources: the client and within your own organization.

    Engineering is almost 2 months behind schedule and construction, although started on schedule, is also behind schedule, by about a month. The contract is lump sum, turnkey, and you are currently forecasting a 10% overrun in the lump sum price.

Problem

1. You have received copies of the attached letters related to requests for additional work, changes, and clarifications. The project manager has asked you to evaluate each of the letters and determine which of the following categories they fall in:

   a. a contractual change order

   b. a change or request within the job scope or contract

   Be prepared to discuss the reasons for your answers.

2. List at least four undesirable side effects on the project from the attached changes and/or requests. How do you plan to account for these side effects in your lump sum estimate for each change?

3. What steps can you recommend that the PM take to minimize the impact of these changes?

4. Based on the current job status and the history of change order requests to date, what steps would you suggest the the PM take to control future changes?

Attached is an excerpt from the lump sum contract agreement which relates to the contractual conditions associated with change orders.

ARTICLE 7—CHANGES

7.1. *Discretionary Rights of Owner*

    7.1.1. OWNER shall have the right, without additional consent from CONTRACTOR, to (i) revise JOB SPECIFICATION, (ii) change elements of WORK already completed or being performed in accordance with JOB SPECIFICATION, or (iii) omit a part of WORK previously authorized, provided such revision or change is within the general scope of WORK specified in CONTRACT on CONTRACT date.

    7.1.2. OWNER shall have the right, without additional consent from CONTRACTOR, to (i) make final decisions on the interpretation of JOB SPECIFICATION and on matters where JOB SPECIFICATION permits alternatives or is not specific, (ii) provide, designate, or reject sources of supply for services, equipment, materials or supplies that JOB SPECIFICATION requires CONTRACTOR to provide, and (iii) require CONTRACTOR to provide engineering studies and cost estimates needed to ascertain the effects of a proposed JOB SPECIFICATION revision.

    7.1.3. At OWNER's request, CONTRACTOR shall furnish under the provisions of CONTRACT additional services that are outside the general scope of WORK specified in CONTRACT on CONTRACT date, provided such additional services are within CONTRACTOR's personnel capacity at the time of OWNER's request.

    7.1.4. A CHANGE ORDER shall be issued with respect to the matters specified in 7.1.1, 7.1.2 and 7.1.3 if appropriate under the criteria of 7.3.1.

7.2. CONTRACTOR shall promptly comply with instructions, authorizations and notices given by OWNER with respect to WORK notwithstanding that a CHANGE ORDER has not been issued or that agreement has not been reached on the effects, if any, on CONTRACT PRICE BUDGET and/or SCHEDULED COMPLETION DATE.

7.3. *Change Orders*

    7.3.1. Unless CONTRACT provides otherwise, OWNER shall issue a CHANGE ORDER when it revises JOB SPECIFICATION or elements of WORK already completed or being performed in accordance with JOB SPECIFICATION, requires additional services of CONTRACTOR or directs omission of part of WORK previously authorized, providing either of the following CHANGE ORDER criteria is satisfied:

        a. CONTRACTOR's costs for performing WORK are affected thereby.

b.   The time required for performing WORK is affected thereby.

7.3.2.   If either of the foregoing criteria is satisfied, OWNER shall authorize CONTRACTOR to prepare and CONTRACTOR shall prepare an estimate of the effects on CONTRACT PRICE BUDGET, SCHEDULED COMPETION DATE. After CONTRACTOR and OWNER agree on the lump sum price, OWNER shall issue a CHANGE ORDER adjusting any or all of the two aforesaid items unless CONTRACT provides otherwise.

7.3.3.   A CHANGE ORDER shall not be allowed when:

a.   documents required to complete the initial issue of JOB SPECIFICATION are issued, unless they conflict with JOB SPECIFICATION at CONTRACT date,

b.   minor design revisions are required to recently completed designs, or while the design work is in progress,

c.   revisions in WORK already performed by CONTRACTOR are required to achieve compliance with JOB SPECIFICATION or to correct errors, omissions, or work not in accordance with sound and generally accepted engineering and construction practices,

d.   studies and cost estimates are required by OWNER to assure optimum design and/or construction,

e.   overtime work is performed or premiums or bonuses are paid by CONTRACTOR for earlier delivery of equipment or materials unless such action (i) is taken at OWNER's request and (ii) did not result, without being excusable under force majeure, from CONTRACTOR's having fallen behind the agreed upon detailed work schedule.

7.3.4.   If, when considering a revision to JOB SPECIFICATION, OWNER authorizes studies or estimates pursuant to 7.1.2 but decides to not proceed with the revision, the CHANGE ORDER issued to cover the study or estimating effort shall not adjust SCHEDULED COMPLETION DATE.

7.3.5.   If CONTRACTOR believes that any instruction, interpretation, decision or other act or omission of OWNER, including unreasonable delays in providing approvals, authorizations, agreements or reviews, meets the criteria for a CHANGE ORDER pursuant to 7.3.1, CONTRACTOR shall promptly notify OWNER thereof. To the extent it reasonably agrees, OWNER shall issue a CHANGE ORDER pursuant to 7.3.2. No CHANGE ORDER shall be allowed if CONTRACTOR has proceeded with the work affected by said act or omission prior to notifying ONWER or if in OWNER'S judgement:

     a.   said act affected CONTRACTOR's performance in a manner consistent with the requirements of CONTRACT or was necessitated by CONTRACTOR's failure to comply with a requirement of CONTRACT, or

     b.   CONTRACTOR's performance was adversely affected by another cause, including CONTRACTOR's fault or negligence.

7.3.6.   In addition to the exercise by OWNER of its descretionary rights set forth in 7.1, certain circumstances are identified elsewhere in PRINCIPAL DOCUMENT (e.g., suspension, termination, force majeure) which specifically require OWNER to issue CHANGE ORDERS to document the effect of such circumstances on CONTRACT PRICE BUDGET and SCHEDULED COMPLETION DATE: When such a circumstance occurs, OWNER shall authorize CONTRACTOR to estimate the circumstance's effect on CONTRACT PRICE BUDGET, SCHEDULED COMPLETION DATE. After CONTRACTOR and OWNER agree on the reasonable effects of the circumstance on these items, OWNER shall issue a CHANGE ORDER adjusting any or all of the two aforesaid items unless CONTRACT provides otherwise.

7.3.7.   CONTRACT PRICE BUDGET and SCHEDULED COMPLETION DATE shall be subject to adjustment only by CHANGE ORDERS, except as otherwise provided in 23.3. A CHANGE ORDER, when issued, shall be deemed to include the effect of the change in WORK or the circumstance covered therein on all previously authorized WORK.

END OF ARTICLE

ARTICLE 26—FORCE MAJEURE

26.1.   No delay or failure in performance by either party hereto shall constitute default hereunder or give rise to any claim for damages if, and to the extent, such delay or failure is caused by force majeure.

26.2.   Force majeure is an occurrence beyond the control and without the fault or negligence of the party affected and which said party is unable to prevent or provide against by the exercise of reasonable diligence including, but not limited to: acts of God or the public enemy; expropriation or confiscation of facilites; changes in applicable LAW; war, rebellion, sabotage or riots, floods, unusually severe weather that could not reasonably have been anticipated; fires, explosions, or other catastrophies; strikes or any other concerted acts of workers; other similar occurrences.

26.3.   The following are specifically excluded as force majeure occurrences unless (i) they were caused by force majeure occurrences of the type set forth in 26.2, and (ii) an acceptable alternate source of services, equipment, or materials is unavailable.

26.3.1. Late performance by a subcontractor caused by a shortage of supervisors or labor, inefficiencies, or similar occurrences.

26.3.2. Late delivery of equipment or materials caused by congestion at a manufacturer's plant or elsewhere, an oversold condition of the market, inefficiencies, or similar occurrences.

26.4. If CONTRACTOR is delayed in performance of WORK by an occurrence it feels is force majeure, CONTRACTOR shall promptly notify OWNER, which if it agrees, shall then notify CONTRACTOR confirming the existence of force majeure. When the effects of said occurrence can be estimated, OWNER shall issue a CHANGE ORDER.

26.5. CONTRACTOR shall make reasonable efforts to minimize the effects of a force majeure occurrence on COSTS and COMPLETION DATE.

END OF ARTICLE

(Owner's Name)	Coordination Procedure
Job Specification	Section 18
(Name) Project	Revision No. ____

(Date)
Page 1801

CHANGES IN WORK

1.0. Scope

This Section outlines the Owner's requirements for processing by the Contractor of change requests.

2.0. Prompt Handling of Changes

Revisions to the Job Specification should be expected during the normal course of engineering, procurement, and construction. The Contractor shall respond promptly to requests for estimates of the effects, if any, that a proposed revision will have on the Contract Price Budget, and Scheduled Completion Date. Contractor shall also respond promptly to requests for estimates of the effects of circumstances, identified in the Principal Document, for which Change Orders are allowable.

3.0. Procedures

3.1. The detailed procedures to be followed in the processing of change requests shall be agreed between the Owner and the Contractor. Those procedures should be compatible with the Contractor's normal methods, including use of standard forms, etc., provided these meet the requirements of this section and establish appropriate, control measures for handling changes.

The procedures; which are to be documented by the Contractor, shall include details on the following:

3.1.1. Initiation of change requests.

3.1.2. Preparation of change proposals and timeliness thereof.

3.1.3. Formalizing and issuance of Change Orders.

3.1.4. Implementing Job Specification revisions.

3.2. To minimize the time required for estimating and to facilitate overall review, the Contractor shall use unit costs data which are consistent with your estimating basis. The use of such unit cost data will also insure that the quality and degree of detail in change proposal estimates parallel those in the estimates used to develop Contract Price Budget. Additionally, agreed percentages shall be used to reflect field labor overhead charges associated with increases/decreases in direct labor.

3.3 The Contractor shall respond to change requests by submitting a change proposal to the Owner within 7 calendar days of the initial request. The proposal shall include:

3.3.1. A brief description of revisions to the Job Specification and/or services involved, including appropriate identifying references.

3.3.2 Effects on Contract Price Budget with cost subtotals reflecting estimated increases/decreases in direct material, direct labor, subcontracts, indirect field costs, home office and branch office costs, freight and duty, other costs (to be identified), fee, and cost of preparing the change proposal. The estimates shall be supported by accompanying backup data which clearly define how subtotals were developed.

3.3.3. Effect on Scheduled Completion Date, if any, with appropriate backup.

3.3.4. An indication of who initiated the change request (Owner, or Contractor) and reason therefore (safety, operability, Owner preference, investment return).

3.3.5. If applicable, a statement on the effect of the change on the Contractor's and Vendors' guarantees.

3.3.6. Effect, if any, on process or utility requirements.

3.4. If authorization to proceed with the revision has not been granted by the Owner, the change proposal shall include a statement defining the latest date such authorization can be given without further affecting the Scheduled Completion Date.

3.5. If the Contractor determines that the 7 calendar day schedule cannot be met, it shall promptly advise the Owner, state the reason for the delay, and the date the change proposal will be available for review.

3.6. Contractor's project manager or authorized representative shall countersign all Change Orders issued by Owner to document Contractor's receipt and agreement.

3.7. Contractor shall maintain a Change Order summary that is to include the following in tabular form: Change Order number, brief description of change, date change proposal is submitted to Owner, date approved or refjected, action on change (approval or rejection), effects on Contract Price Budget, Scheduled Completion Date and remarks.

3.8. Each change request shall be assigned a Change Order number and be entered on the summary at the time it is requested. Subsequent entries shall be made at appropriate times to ensure that the summary is current. The Contractor shall issue the summary to the Owner monthly with its progress report and, otherwise, as requested by the Owner.

4.0. Work Authorization

4.1. Authorization by the Owner, for the Contractor to perform work associated with a change, will normally accompany the approved Change Order.

4.2. At the Owner's discretion, Owner may authorize the Contractor to perform all or part of the work associated with the change (in addition to preparation of a change proposal) at the time of issuing a change request.

4.3. If Contractor wishes to initiate a change request, he shall obtain Owner's approval before expending any engineering, cost estimating, scheduling or other effort in support of the request. The Owner if he approves, will issue a change request.

5.0. Distribution of Change Documents

Change proposals and Change Orders shall be distributed in accordance with Section 2, Schedule C.

CHARACTERS

Monaco Encon (the owner)

Tom Sharpshooter, project executive	Bill Perkins, project engineer
Harry Head, startup leader	Joe Turnbuckle, safety engineer

We Can Do Contractors (the contractor)

John Doe, project manager	Hi Reading, instrument specialist

Mr. John Q. Doe
Project Manager
c/o We Can Do Contractors
2380 Borderline Avenue
Houston, Texas 77840

PMT - 1
October 14, Year 1
Subject: Nozzle Access Platform

Dear John:

    During a recent plant inspection, John Smith, maintenance superintendent, suggested a platform should be installed below Nozzle N-6 on tower T-101 to permit blinding the 10 in. flange at nozzle N-6. This would allow line Ala-1006 to be bled prior to the removal of pump P-106. Would you please proceed with the necessary steps to accomplish this installation.

Sincerely,

                   Harry B. Head
                   Startup Team Leader
                   Monaco Encon, Inc.

Mr. Tom Sharpshooter
Project Executive
Monaco Encon, Inc.
Monte Carlo, Monaco

PMT - 2
October 28, Year 1
Subject: Relocation of TI-201

Dear Tom:

    Resulting from your review of our Process and Instrument Diagram, the location of TI-201 is to be changed so it will measure the temperature downstream of the mixpoint of branch Alc-3024. The relocation will cost nothing extra in piping or in other material and labor. However, we do request that a change be issued for the extra engineering hours expended.

Sincerely,

John Q. Doe
WE CAN DO CONTRACTORS
Project Manager

Note: Upon reexamination of original Design Flow Plan No. S1-2 by the PM Team, TI-201 is shown to be located downstream of the mixpoint of branch Alc-3024.

Mr. John Q. Doe
Project Manager
c/o We Can Do Contractors
2380 Borderline Avenue
Houston, Texas 77840

PMT - 3
November 7, Year 1
Subject: Addition of Water Pump
and Water Heater

Dear John:

During a recent meeting with Plant Operations personnel, the addition of an industrial water pump to boost the water pressure to the chlorinators and the addition of an industrial water heater for the furnace preheater was requested. The additions will facilitate satisfactory unit maintenance/housekeeping. Would you please indicate to me the extent to which these additions will affect cost and schedule.

Sincerely,

Bill T. Perkins
Project Engineer
Monaco Encon, Inc.

Mr. Tom Sharpshooter
Project Executive
Monaco Encon, Inc.
Monte Carlo, Monaco

PMT - 4
October 26, Year 1
Subject: Jobsite Shutdown

Dear Tom:

As you know, our jobsite was shut down yesterday, October 25, due to a walkout by the ironworkers. The picket line was honored and all work on the jobsite subsequently ceased. This serves to notify you that we are claiming Force Majeure under the terms of the contract. We will advise you of the cost and schedule impact to the project as soon as our assessment is complete.

Sincerely,

John Q. Doe
WE CAN DO CONTRACTORS
Project Manager

Mr. John Q. Doe
Project Manager
c/o We Can Do Contractors
2380 Borderline Ave.
Houston, Texas 77840

PMT - 5
January 10th, Year 2

Dear John:

As part of the furnace burner replacement on the Visfiner some repaving is to be done to refurbish the area after realignment of air ducts. It has been suggested to me that the whole access way into that area is a safety hazard due to the break-up of the surface of the access way (approx. 40 m × 10 m). Can you please arrange for repaving of this accessway to be added to the repaving subcontract already included in the project.

Very truly yours,

Tom Sharpshooter
Project Executive
Monaco Encon, Inc.

Mr. Tom Sharpshooter
Project Executive
Monaco Encon, Inc.
Monte Carlo, Monaco

PMT - 6
May 20 Year 1

Change List - 2

Dear Tom:

    It is now one month since we received Change List 2 to the Visfiner Design Specification. To avoid delay in the project we have, as requested in your transmittal letter, incorporated CL-2 into our work. We have not yet received a change request to properly add this Change List to the project. We request that this matter be quickly remedied.

Very truly yours,

John Q. Doe
Project Manager

Mr. Tom Sharpshooter
Project Executive
Monaco Encon, Inc.
Monte Carlo, Monaco

PMT - 7
April 1, Year 1

Furnace Burner Refractory

Dear Tom,

    In the most recent approved vendors list the type and supplier of refractory for the new furnace burners. have been changed. We have thus been obliged to issue new documentation to the bidders for the burners. Can you please issue a change request to cover this change to the Job Specification.

Very truly yours,

John Q. Doe
Project Manager

Mr. John Q. Doe
Project Manager
c/o We Can Do Contractors
2380 Borderline Ave.
Houston, Texas 77840

PMT - 8
September 3rd, Year 1

*Emergency Exit
from Control House*

Dear John:

As the result of model review, it has been suggested to me by the operations department that a safety exit from the control house be made by erecting a low walkway bridge over the grade level pipe way between the back door of the control house and avenue '3'. I support such a safety measure. Can you therefore include such a walkway bridge over the pipe rack.

Very truly yours,

Bill T. Perkins
Project Engineer
Monaco Encon, Inc.

Note: A call to Bill Dozer We Can Do field engineer confirms that there is an accessway 7 m away from the back door of the control house which gives access onto avenue '3'. The insulation on the pipes in the pipeway opposite the back door is however flattened where operators have been using the pipeway to the refinery bus pick-up point.

Mr. John Q. Doe
Project Manager
c/o We Can Do Contractors
2380 Borderline Avenue
Houston, Texas 77840

PMT - 9
October 24, Year 1
Subject: Hydrogen Unit

Dear John:

As you know, during our recent meeting with you and your process follow-up engineers we concluded that it would be necessary to change the hydrogen unit from a single train per DS 81-103 to double trains, thus permitting continuous operation. Each train should be sized for sixty percent operational capacity and should be piped to permit simultaneous or separate operation. It is my understanding that you will pursue the ramifications of this alteration.

Sincerely,

Tom Sharpshooter
Project Executive
Monaco Encon, Inc.

Mr. John Q. Doe
Project Manager
c/o We Can Do Contractors
2380 Borderline Avenue
Houston, Texas 77840

PMT - 10
October 29, Year 1
Subject: Start-up Line

Dear John:

To allow the start-up to progress as planned, I recommend installing a temporary start-up line between fuel gas line FG Ala-1234 and Furnace F-101. This line is to permit furnace dryout prior to start-up. Would you please proceed with the necessary steps to accomplish this installation.

Sincerely,

Harry B. Head
Start-up Team Leader
Monaco Encon, Inc.

Mr. John Q. Doe
Project Manager
c/o We Can Do Contractors
2380 Borderline Avenue
Houston, Texas 77840

PMT - 11
October 20, Year 1
Subject: Block and Bypass
Valves at CV-106

Dear John:

The nature of the slurry in line A3b-3026 is such that it will erode the valve seat at CV-106 necessitating periodic repair. To prevent unit shutdown during repair of CV-106, I recommend the addition of block and bypass valves at CV-106. Would you issue the necessary documentation for this addition.

Sincerely,

Hi Reading
Instrument Specialist
We Can Do Contractors

Mr. John Q. Doe
Project Manager
c/o We Can Do Contractors
2380 Borderline Avenue
Houston, Texas 77840

PMT - 12
October 31, Year 1
Subject: Additional Tank

Dear John:

As a result of changing market conditions, there exists a need for an additional tank of the same type as TK-102. This tank would not only double the capacity now indicated by DS 81-103, but would also increase our operational flexibility. Would you pursue the cost and schedule ramifications of this addition.

Sincerely,

Tom Sharpshooter
Project Executive
Monaco Encon, Inc.

Mr. John Q. Doe
Project Manager
c/o We Can Do Contractors
2380 Borderline Avenue
Houston, Texas 77840

PMT - 13
November 4, Year 1
Subject: New NPQC Safety
Requirement

Dear John:

As part of the safety audit conducted last week, Greg Falls pointed out the
new safety requirement that specifies the use of foam glass insulation in areas
that are susceptible to fuel oil spillage. The following areas are affected by this
requirement:

Fuel Oil Strainers	PV-3	STR-119A, B
	G-02	STR-2A, B
Bottom Pump Strainers	PV-3	STR-110A, B, C
Front of SP-4 burners		
Vicinity of fuel oil sample outlets		

Would you take the necessary action to comply with this safety requirement.

Sincerely,

Joe Turnbuckle
Safety Engineer
Monaco Encon, Inc.

CASE STUDY NO. 9: CONSTRUCTION PROGRESS MEASURMENT AND PRODUCTIVITY ANALYSIS: STATUS OF CONCRETE CONSTRUCTION-TIJUANA PROJECT

The purpose of this case study is to develop the student's understanding of the principles discussed in Chap. 21. It is a real-life situation that is frequently encountered by cost engineers assigned to construction management teams.

Scenario

You are Lumor's field cost engineer working on a fully reimbursable project for the Tijuana Oil Company. Today is the 5th of December and the construction manager has just stopped in your office and advised you that Tijuana's project manager has requested that a complete review on the status of concrete as of end of November be presented tomorrow morning at the monthly cost and progress meeting The construction manager would like the following information available for a preliminary review by quitting time tonight.

1. End of November overall percent complete for concrete.
2. Labor productivity during November and cumulative to date.
3. Forecasted total manhours to complete all concrete work.
4. Any particular subaccounts that are in trouble and tending to drag down overall productivity.
5. Estimate increased cost if Tijuana decides to put all concrete crews on a 60 hr week to improve the schedule.

Information Available

1. The quantity adjusted manhour budget for each concrete subaccount taken from November's computerized manpower report has been copied onto the attached work sheet.
2. Change order #302 adding additional paving has been approved but not yet input into the computerized QAB. The detailed estimate for CO #302 shows 2400 manhours in the paving account.
3. Current measured percents complete for each subaccount taken from the computer report have been copied onto the attached work sheet. The paving percent complete reflects CO #302.
4. Workhours used to end of November obtained from the computer reports have been copied onto the attached work sheet.
5. Lumor's cost/scheduling manual does not have productivity profile calibration curves for the concrete subaccounts, but it does have a calibration curve for the overall concrete account. The profile which you have been plotting is attached. Both period and cumulative productivity have been plotted through October. November's data has not yet been plotted.

6.   The current average labor wage rate for the concrete account is $9.50 per manhour. This figure includes an average of 2% spot overtime calculated on an hours worked basis. Time and one-half is paid for overtime by union agreement.

7.   Lumor's cost/scheduling manual includes a curve indicating that working 60 hr weeks over a prolonged period will cause an average productivity drop of 20% compared to productivity achieved by working normal 40 hr weeks with less than 3% spot overtime.

8.   A short discussion with Red-Mix, the concrete craft superintendent, reveals that his concrete pump, used for pouring concrete into the elevated forms for structural concrete, has been broken for the past 6 weeks. Red says that maintenance has not advised him as to what is holding up repair of the pump, but he has rigged up a temporary hoist and obtained some large buckets so that work can continue until the pump is repaired. A quick check of weekly productivity reports shows that productivity in the structural subaccount dropped off about 10% six weeks ago and it has not improved since.

9.   A phone call to the head of equipment maintenance reveals that a spare part needed for repair of the concrete pump was ordered 3 weeks ago but had not yet arrived. No, maintenance had not followed up with the spare part vendor as to why the part had not arrived. Nobody advised that the concrete pump was urgently needed. Yes, the spare part could be fabricated in 24 hrs by the local machine shop if someone authorized overtime and expenditure of the extra money.

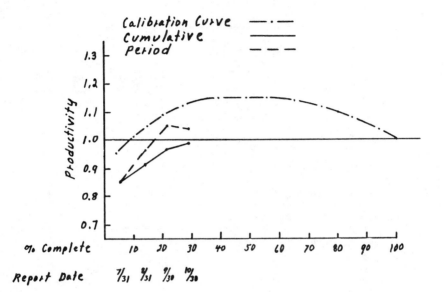

PROBLEM N° 7

LUMOR ENGINEERS & CONSTRUCTORS
WORKSHEET FOR DIRECT LABOR ANALYSIS

Account: 30-XX Concrete

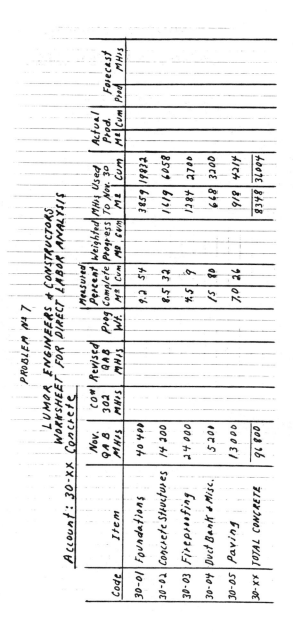

Code	Item	Nov. QAB MHrs	C.O.# 302 MHrs	Revised QAB MHrs	Prog Wt.	Measured Percent Complete M°	Cum	Weighted Progress M° Cum	Cum	MHrs Used To Nov. 30 M°	Cum	Actual Prod. M° Cum	Forecast Prod MHrs
30-01	Foundations	40400				9.2	54			3859	19832		
30-02	Concrete Structures	14200				8.5	32			1619	6058		
30-03	Fireproofing	24000				4.5	9			1284	2700		
30-04	Duct Bank & Misc.	5200				15	80			668	3200		
30-05	Paving	13000				7.0	26			918	4214		
30-XX	TOTAL CONCRETE	96800								8348	36004		

CASE STUDY NO. 10: CONSTRUCTION PROGRESS
MEASUREMENT AND LABOR FORECAST: EXPANSION
OF MONACO-ENCON FACILITIES BY LUMOR

This case study is designed to expand the students' understanding of construction progress measurement, labor productivity, and manpower forecasting. It depicts a complex situation encountered on an actual project. Solution of this study should generate a considerable amount of discussion, and it is recommended that students be divided into groups of two or four persons, and undertake this study after reading Chap. 21.

Scenario

You are the lead cost engineer on Lumor's construction management staff working on a large grassroots project for Monaco Encon. Your construction manager, aware of labor problems in the atmosphere pipestill unit, has asked you for a complete assessment of the labor outlook in this area. He wants to be brought up to date on the percent complete, productivity problems, and forecasted total manhours to complete all work in the area.

Construction work is approaching mechanical completion in the atmospheric pipestill (APS) unit. Attached is a work sheet which is used by Lumor to calculate their monthly physical progress for the unit.

Problem

In order to fully inform the construction manager you should do the following:

1. Complete the work sheet through column E and determine the physical percent complete for the APS unit for this month.
2. Calculate the producitivity to date and enter data in column G.
3. List the factors which will influence productivity during remainder of the project schedule and which must be considered in arriving at the forecasted productivity and forecasted manhours.
4. Is the forecast productivity most often less or more than the productivity to date? Why?
5. Considering preceding points 3 and 4 plus the attached productivity profiles, forecast final job productivity and total manhours required for project completion. Enter these forecasts in columns H and I of the work sheet.
6. What are the major steps which must be taken to control productivity in the latter stages of a project?
7. Recommend corrective action that can be taken immediately to improve productivity and that you have reflected in your manhour forecast.

Additional Information

1. Productivity profiles with calibration curves based on 3 previous Lumor refinery projects are attached. Also, actual productivity as experienced on the Monaco Encon project is plotted on the same profiles.
2. Above-ground piping erection is about 3 weeks behind schedule; therefore, piping erection crews have worked every Saturday for the past month (i.e., 48 hrs per week).
3. A shortage of skilled welders has been identified as a bottleneck in piping erection. Intensive recruiting in the local area has failed to locate the qualified pressure piping welders; however, some structural welders are available and a number of applicants supplied by the union exhibited some ability but failed the piping qualification tests.
4. Starting next Monday the Lumor construction manager plans to put all piping welders on a 60 hr week working 6 10-hr days.
5. First class pipe fitters are available for hiring at the union hall.
6. The piping superintendent has recently had bitter quarrels with both the electrical and instrument superintendents because his people have, allegedly, through carelessness, broken an installed conduit and smashed a number of instruments.
7. The piping superintendent claims that he is continuously delayed by lack of scaffolding; however, the scaffolding subcontractor can prove that he has executed all work orders within the scheduled time limit.
8. A sampling study by the cost engineers points out that piping personnel are losing a great deal of time helping the warehouse personnel locate required fabricated piping spools. Spools have been unloaded in piles and not sorted by the warehouse personnel. The piping superintendent has never mentioned this problem.

TYPICAL WORK SHEET: PHYSICAL PROGRESS CALCULATION

PRIMARY CODE APS UNIT

Secondary Code & Description	(A) Manhours control estimate	(B) Weight factor	(C) % Physical Progress last month	(D) % Physical Progress this month	(E) Weighted % complete	(F) Manhours expended	(G) Prod. to date	(H) Forecast manhours	(I) Forecast prod.	Notes
0XXX Earthwork[a]	38160		100	100		41934				(A) Column A reflects the original estimate plus all change orders and quantity adjustments. It is the current QAB.
11XX Concrete found.	101500		100	100		109140				
17XX Fireproof & Grout	62800		100	100		69780				
19XX Paving	16300		100	100		11810				
2XXX Str. steel	50900		100	100		53579				
42XX Shop fab. vessels	16800		100	100		16000				
43XX Compressors	5200		70	95		5312				(B) Weight Factor
44XX Heat exchangers	29500		80	90		28548				(C) From last month's report.
45XX Furnaces	74000		100	100		80435				(D) Reported *Physical Percent complete* for each code.
46XX Pumps	11200		100	100		11205				
56XX Pipe U.G.	6200		100	100		6301				(E)
57XX Pipe A.G.	235900		43	58		145555				(F) Firm field manhour reports of actual expenditures.
60XX Elect. U.G.	14700		100	100		15978				
65XX Elect. A.G.	56300		50	70		40629				(G)
7XXX Instruments	66200		30	50		34479				(H) From Productivity Review Forecast.
83XX Paint	19300		15	20		3939				(I) Actually determined by analyzing secondary accounts in each unit.
85XX Insulation	66200		30	40		27020				Does not include indirect accounts...capital account only.
TOTAL	871160					701644				

[a] Contractor's Code of Accounts
(Typical worksheet for one unit only)
NOTE: PM may or may not want to have this detail in the montly progress reports.

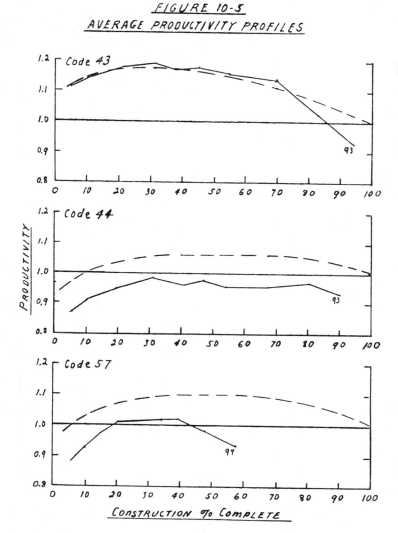

FIGURE 10-5
AVERAGE PRODUCTIVITY PROFILES

Calibration Basis: 3 Previous Lumor Refinery Projects

Legend:

Calibration Curve   — — — —

Actual Cummulative Monaco Encon Project   ———————

R65

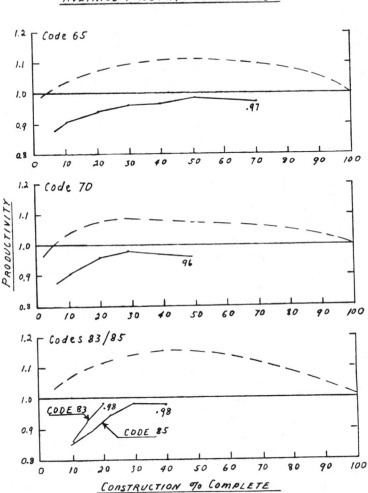

FIGURE 10-6

AVERAGE PRODUCTIVITY PROFILES

CONSTRUCTION % COMPLETE

<u>Calibration Basis</u> : 3 Previous Lumor Refinery Projects

<u>Legend</u> :

Calibration Curve   — — — —

Actual Cummulative Monaco Encon Project   ————

INDEX

INDEX